Macro and Micro Univer

Preon Interaction Theory and Model of Universe

Alexander Bolonkin

USA, Lulu, 2017

Title: Preon Interaction Theory and Model of Universe
Author: Alexander Bolonkin, abolonkin@gmail.com
ISBN: 978-1-365-79994-6

Book contains researches six new ideas: new preon interaction theory of the micro World; relations between time, mass, space, charge and energy; possibility of creating the super-strong (in millions times) matter, having suprice properties; super-strong nuclear AB-needles, which allows to penetrate deep into the Earth and planets; the nuclear generator for converting of any matter into the energy and possibility of the artificial explosion of Sun.

Copyright @2017 by author-Lulu

Published Lulu in USA, www.lulu.com .

Contents

About author

Abstract

Chapter 1. Preon Interaction Theory and Model of Universe. 7

Chapter 2. Relations between Time, Mass, Space, Charge and Energy (part 3). 19

Chapter 3. Super-strong nuclear matter. 27

Chapter 4. AB needles. Stability and production super-strong nuclear matter. 45

Chapter 5. Converting the matter in energy. 61

Chapter 6. Artificial Explosion of Sun and Criterion for Solar Denonation 77

Appendix 100

About the Author

Bolonkin, Alexander Alexandrovich (1933-)

Alexander A. Bolonkin was born in the former USSR. He holds doctoral degree in aviation engineering from Moscow Aviation Institute and a post-doctoral degree in aerospace engineering from Leningrad Polytechnic University. He has held the positions of senior engineer in the Antonov Aircraft Design Company and Chairman of the Reliability Department in the Clushko Rocket Design Company. He has also lectured at the Moscow Aviation Universities. Following his arrival in the United States in 1988, he lectured at the New Jersey Institute of Technology and worked as a Senior Researcher at NASA and the US Air Force Research Laboratories.

Bolonkin is the author of more than 250 (2015) scientific articles and books and has 17 inventions to his credit. His most notable books include The Development of Soviet Rocket Engines (Delphic Ass., Inc., Washington , 1991); Non-Rocket Space Launch and Flight (Elsevier, 2006); New Concepts, Ideas, Innovation in Aerospace, Technology and Human Life (NOVA, 2007); Macro-Projects: Environment and Technology (NOVA, 2008); Human Immortality and Electronic Civilization, 3-rd Edition, (Lulu, 2007; Publish America, 2010); Femtotechnologies and Revolutionary Projects, LAMBERT, 2011; Innovations and New Technologies (v.2), Lulu, 2013; Life and Science, LAMBERT, 2011; etc. .

Abstract

Book contains researches five new ideas: new preon interaction theory of the micro World; relations between time, mass, space, charge and energy; possibility of creating the super-strong (in millions times) matter, having suprice properties; super-strong nuclear AB-needles, which allows to penetrate deep into the Earth and planets; the nuclear geterator that is converting of any matter into energy.

1. In Chapter 1 the author offers some initial ideas about a cognitive construct of the Micro-World with allows to design a preon based Universe matching many qualities of the observable Universe. The main idea is that - the initial base must be very simple: two energy massless virtual particles (eners) and two reciprocity relations (interactions) between them. Author postulates: Two energy massless virtual particles can explain the main features of much of what we see including: mass, electrical charges and the main interactions between particles such as: gravitation, centrifugal and inertial masses, repulsion and attraction of electric charges, weak and strong nuclear forces, design of quarks and baryonic matter.

2. In Chapter 2 author has developed a theory which allows derivation of the unknown relations between the main parameters (energy, time, volume, matter) in the Universe. In given part 3 he added charge as main parameter in this theory. He finds also the quantum (minimal values) of energy, time, volume and matter and he applied these quantum for estimations of quantum volatility and the estimation of some values of our Universe and received both well-known and new unknown relations.

Author offers possibly valid relations between charge, time, matter, volume, distance, and energy. The net picture derived is that in the Universe exists ONLY one substance – ENERGY. Charge, time, matter, volume, fields are evidence of this energy and they can be transformed one to other. Author gives the equations which allow to calculate these transformation like the famous formula $E = mc^2$. Some assumptions about the structure of the Universe follow from these relations.

Most offered equations give results close to approximately known data of Universe, the others allow checking up by experiment.

3. In Chapter 3 the author researches the design the super-strong matter. This matter is stronger than convetional mathriales in millions times. It is can withstand temperatures in millions degree. Aerospace, aviation particularly need, in any era, the strongest and most thermostable materials available, often at nearly any price. The Space Elevator, space ships (especially during atmospheric reentry), rocket combustion chambers, thermally challenged engine surfaces, hypersonic aircraft materials better than any now available, with undreamed of performance as the reward if obtained. As it is shown in this research, the offered new material allows greatly to improve the all characteristics of space ships, rockets, engines and aircraft and design new types space, propulsion, aviation systems.

At present the term 'nanotechnology' is well known – in its' ideal form, the flawless and completely controlled design of conventional molecular matter from molecules or atoms.

But even this yet unachieved goal is not the end of material science possibilities. The author herein offers the idea of design of new forms of nuclear matter from nucleons (neutrons, protons), electrons, and other nuclear particles. He shows this new nuclear matter has extraordinary properties (for example, tensile strength, stiffness, hardness, critical temperature, superconductivity, supertransparency, zero friction, etc.), which are up to millions of times better than corresponding properties of conventional molecular matter. He shows concepts of design for space ships, rockets, aircraft, sea ships, transportation, thermonuclear reactors, constructions, and so on from nuclear matter. These vehicles will have unbelievable possibilities (e.g., invisibility, ghost-like penetration through any walls and armour, protection from nuclear bomb explosions and any radiation flux, etc.

Nanotechnology, in near term prospect, operates with objects (molecules and atoms) having the size in nanometer (10^{-9} m). The author here outlines perhaps more distant operations with objects (nuclei) having size in the femtometer range, (10^{-15} m, millions of times less small than the nanometer scale). The name of this new technology is femtotechnology.

4. In Chapter 4 author offered and considered possible super strong nuclear matter. In given work he continues to study the problem of a stability and production this matter. He shows the special artificial forms of nuclear AB-matter which make its stability and give the fantastic properties. For example, by the offered AB-needle you can pierce any body without any damage, support motionless satellite, reach the other planet, and research Earth's interior. These forms of nuclear matter are not in nature now, and nanotubes are also not in nature. The AB-matter is also not natural now, but researching and investigating their possibility, properties, stability and production are necessary for creating them.

5. In Chapter 5 Author offers a new nuclear generator which allows to convert any matter to nuclear energy in accordance with the Einstein equation $E=mc^2$. The method is based upon tapping the energy potential of a Micro Black Hole (MBH) and the Hawking radiation created by this MBH. As is well-known, the vacuum continuously produces virtual pairs of particles and antiparticles, in particular, the photons and anti-photons. The MBH event horizon allows separating them. Anti-photons can be moved to the MBH and be annihilated; decreasing the mass of the MBH, the resulting photons leave the MBH neighborhood as Hawking radiation. The offered nuclear generator (named by author as AB-Generator) utilizes the Hawking radiation and injects the matter into MBH and keeps MBH in a stable state with near-constant mass. The AB-Generator can not only produce gigantic energy outputs but should be hundreds of times cheaper than a conventional electric generation processes. The AB-Generator can be used in aerospace as a photon rocket or as a power source for numerous space vehicles. Many scientists expect the Large Hadron Collider at CERN will produce one MBH every second and the technology to capture them may be used for the AB-Generator.

Chapter 1
Preon Interaction Theory and Model of Universe

When God created the World, he did not know
of string theory or quantum mechanics.
He used the Principle of Simplicity.

Abstract

Author offers some initial ideas about a cognitive construct of the Micro-World with allows to design a preon based Universe matching many qualities of the observable universe. The main idea is that - the initial base must be very simple: two energy massless virtual particles (eners) and two reciprocity relations (interactions) between them. Author postulates: Two energy massless virtual particles can explain the main features of much of what we see including: mass, electrical charges and the main interactions between particles such as: gravitation, centrifugal and inertial masses, repulsion and attraction of electric charges, weak and strong nuclear forces, design of quarks and baryonic matter.

Author gives only ideas of how these problems may be solved. Scientists who will be interested in the offered approach can make detailed mathematical descriptions and solutions.

Key words: microworld, preon, preon theory, virtual particles, fundamental interactions, Ener Model of Universe, Bolonkin.

Introduction
Short information, discription and history of problems.

Univese.

The Universe is all of time and space and its contents. The Universe includes planets, stars, galaxies, the contents of intergalactic space, the smallest subatomic particles, and all matter and energy. The *observable universe* is about 28 billion parsecs (91 billion light-years) in diameter at the present time. The size of the whole Universe is not known. Observations and the development of physical theories have led to inferences about the composition and evolution of the Universe. At present time the scientist suggest the University contants the Ordinary (baryonic) matter (4.9%), dark matter (26.8%), dark energy (68.3%).

Ordinary matter is at least 10^{53} kg, avarage density is 4.5×10^{-31} g/cm^3, avarage temperture is 2.72548 K.

Observations in the late 1990s indicated the rate of the expansion of the Universe is increasing indicating that the majority of energy is most likely in an unknown form called dark energy. The majority of mass in the universe also appears to exist in an unknown form, called dark matter.

The Big Bang theory is the prevailing cosmological model describing the development of the Universe. Space and time were created in the Big Bang, and these were imbued with a fixed amount of energy and matter; as space expands, the density of that matter and energy decreases. After the initial expansion, the Universe cooled sufficiently to allow the formation first of subatomic particles and later of simple atoms. Giant clouds of these primordial elements later coalesced through gravity to form stars. Assuming that the prevailing model is correct, the age of the Universe is measured to be 13.799±0.021 billion years. There is a lot of speculative model of Universes.

The remaining 4.9% of the mass—energy of our Universe is ordinary matter, that is, atoms, ions, electrons and the objects they form. This matter includes stars, which produce nearly all of the light we see from galaxies, as well as interstellar gas in the interstellar and intergalactic media, planets, and all the objects from everyday life that we can bump into, touch or squeeze.

Of the four fundamental interactions, gravitation is dominant at cosmological length scales, including galaxies and larger-scale structures. Gravity's effects are cumulative; by contrast, the effects of positive and negative charges tend to cancel one another, making electromagnetism relatively insignificant on cosmological length scales. The remaining two interactions, the weak and strong nuclear forces, decline very rapidly with distance; their effects are confined mainly to sub-atomic length scales.

Ordinary matter of our University is composed of two types of particles: quarks and leptons. For example, the proton is formed of two up quarks and one down quark; the neutron is formed of two down quarks and one up quark; and the electron is a kind of lepton.

Ordinary matter and the forces that act on matter can be described in terms of elementary particles. These particles are sometimes described as being fundamental, since they have an unknown substructure, and it is unknown whether or not they are composed of smaller and even more fundamental particles. Of central importance is the Standard Model, a theory that is concerned with electromagnetic interactions and the weak and strong nuclear interactions.[93] The Standard Model is supported by the experimental confirmation of the existence of particles that compose matter: quarks and leptons, and their corresponding "antimatter" duals, as well as the force particles that mediate interactions: the photon, the W and Z bosons, Higgs boson, and the gluon. The Standard Model does not, however, accommodate gravity.

Virtual particles.

In physics, a **virtual particle** is an explanatory conceptual entity that is found in mathematical calculations about quantum field theory. It refers to mathematical terms that have some appearance of representing particles inside a subatomic process such as a collision.

Often the virtual-particle virtual "events" appear to occur close to one another in time, for example within the time scale of a collision, so that they are virtually and apparently "short-lived". It restricts itself to what is actually observable and detectable. Virtual particles are conceptual devices that in a sense try to by-pass Heisenberg's insight, by offering putative or virtual explanatory visualizations for the inner workings of subatomic processes.

The range of forces carried by virtual particles is limited by the uncertainty principle, which regards energy and time as conjugate variables; thus, virtual particles of larger mass have more limited range.

They are "temporary" in the sense that they appear in calculations, but are not detected as single particles. Thus, in mathematical terms, they never appear as indices to the scattering matrix, which is to say, they never appear as the observable inputs and outputs of the physical process being modelled.

There are many observable physical phenomena that arise in interactions involving virtual particles. For bosonic particles that exhibit rest mass when they are free and actual, virtual interactions are characterized by the relatively short range of the force interaction produced by particle exchange. Examples of such short-range interactions are the strong and weak forces, and their associated field bosons. For the gravitational and electromagnetic forces, the zero rest-mass of the associated boson particle permits long-range forces to be mediated by virtual particles.

Elementary particles.

In particle physics, an **elementary particle** or **fundamental particle** is a particle whose substructure is unknown, thus it is unknown whether it is composed of other particles.[1] Known elementary particles include the fundamental fermions (quarks, leptons, antiquarks, and antileptons), which generally are "matter particles" and "antimatter particles", as well as the fundamental bosons (gauge bosons and Higgs boson), which generally are "force particles" that mediate interactions among fermions.[1] A particle containing two or more elementary particles is a *composite particle*.

Via quantum theory, protons and neutrons were found to contain quarks—up quarks and down quarks—now considered elementary particles

Other estimates imply that roughly 10^{97} elementary particles exist in the visible universe (not including dark matter), mostly photons, gravitons, and other massless force carriers.

Fundamental Interactions.

Fundamental interactions, also known as fundamental forces, are the interactions in physical systems that do not appear to be reducible to more basic interactions. There are four conventionally accepted fundamental interactions—gravitational, electromagnetic, strong nuclear, and weak nuclear. Each one is understood as the dynamics of a *field*. The gravitational force is modelled as a continuous classical field. The other three are each modelled as discrete quantum fields, and exhibit a measurable unit or *elementary particle*.

The two nuclear interactions produce strong forces at minuscule, subatomic distances. The strong nuclear interaction is responsible for the binding of atomic nuclei. The weak nuclear interaction also acts on the nucleus, mediating radioactive decay. Electromagnetism and gravity produce significant forces at macroscopic scales where the effects can be seen directly in every day life. Electrical and magnetic fields tend to cancel each other out when large collections of objects are considered, so over the largest distances (on the scale of planets and galaxies), gravity tends to be the dominant force.

Currently the electromagnetic, strong, and weak interactions associate with elementary particles, The electromagnetic force are transferring the photons. The electromagnetic interaction carries are was modelled with the weak interaction, whose force carriers are W and Z bosons, traversing the minuscule distance, in electroweak theory (EWT). Strong nuclear force carriers are gluons, gravitation force carriers are gravitons, electro-magnetic force carriers are photons.

Bosons always carries energy and momentum between the fermions. Currently the theory implies the following power transmission mechanism between the particles: the particle know about other similar particle, produces a carrier that repels (the law of conservation of momentum) of the precursor particles, moving to another particle, and pushes her (transmits its pulses). Other particle acts is similarly [Hawking S., A Brief History of Time. Russin translation, Moscow 2015, p. 91].

Preons.

In particle physics, preons are "point-like" particles, conceived to be subcomponents of quarks and leptons.[1] The word was coined by Jogesh Pati and Abdus Salam in 1974. Interest in preon models peaked in the 1980s but has slowed as the Standard Model of particle physics continues to describe the physics mostly successfully.

Preon theory is motivated by a desire to replicate the achievements of the periodic table, and the later Standard Model which named the "particle zoo", by finding more fundamental answers to the huge number of arbitrary constants present in the Standard Model. It is one of several models to have been put forward in an attempt to provide a more fundamental explanation of the results in experimental and theoretical particle physics. The preon model has attracted comparatively little interest to date among the particle physics community.

The existed preon researches are motivated by the desire to explain already known facts (retrodiction), which include:
1) To reduce the large number of particles, many that differ only in charge, to
   a smaller number of more fundamental particles.
2) To reduce the number of experimental input parameters required by the
   Standard Model.
3) To provide reasons for the very large differences in energy-masses observed
   in supposedly fundamental particles, from the electron neutrino to the top
   quark.
4) To account for neutrino oscillation and mass.
5) The desire to make new nontrivial predictions, for example, to provide
   possible cold dark matter candidates.
6) To explain why there exists only the observed variety of particle species and
   not something else and to reproduce only these observed particles (since the
   prediction of non-observed particles is one of the major theoretical
   problems, as, for example, with supersymmetry).

There are a lot of preons models. The **Rishon model** (RM)[2 – 4] is the most popular and illustrates some of the typical efforts in the field.

The model has two kinds of fundamental particles called **rishons** (which means "primary" in Hebrew). They are **T** ("Third" since it has an electric charge of ⅓ e, or Tohu which means "unformed" in Hebrew Genesis) and **V** ("Vanishes", since it is electrically neutral, or Vohu which means "void" in Hebrew Genesis). All leptons and all flavours of quarks are three-rishon ordered triplets. These groups of three rishons have spin -½.They are as follows:

TTT = antielectron;
VVV = electron neutrino;

TTV, TVT and VTT = three colours of up quarks;
TVV, VTV and VVT = three colours of down antiquarks.
  Each *rishon* has a corresponding antiparticle.
Matter and antimatter are equally abundant in nature in the RM.
Higher generation leptons and quarks are presumed to be excited states of first generation leptons and quarks.
Mass is not explained.
  In the expanded Harari–Seiberg version [2] the rishons possess color and hypercolor, explaining why the only composites are the observed quarks and leptons. Under certain assumptions, it is possible to show that the model allows exactly for three generations of quarks and leptons.

The basic ideas of the offered preon model

Virtual elementary fundamental particles and their features

In our model, we put first the principle of parsimony. We take only two elementary fundamental virtual particles named "A" and "B" (or +, -). They are *massless* (or they have a very small mass not measured by current devices. That mass may be equivalent to binding energy $m = E/c^2$, where E is fluctuation of energy. But they have equal the module of energy. Particle A has *positive energy*; particle B has *negative energy*). We name them the *positive and negative "eners"*. The different particles **attract** one to other (A to B, B to A), the same particles **repel** one to other (A from A, B from B). Vacuum produces pairs of A - B in equal amounts,
create mixture and not require energy for producing because **sum of their energy and momentum equal zero.**

  Their energy may be in form of kinetic or/and rotation. No violations of laws of the conservation of energy, momentum and angular momentum occur.

  Positive and negative eners are *not* conventional particles and antiparticles used in current science. In current science the particles and antiparticles produce a huge energy (conventional in radiation form) when they annihilate. The eners destroy energy, convert it to zero, when they annihilate.

  The eners have a size and produce field and the space where they are located because the any particles have size (space). They can transmit the information because they can have collisions, vibrations and waves, but maximal speed of transmission is limited about c = 300 thousands km/s ($c = 3 \cdot 10^8$ m/s - light speed in vacuum).

  Our word "ener field" has difference significance from the common sense of science. That is a space filled by eners, which can be used as construction material for real particles and fundamental interaction between the produced particles and transfer of interaction (forces).

  The virtual particles (eners) and their initial interaction are shown in fig.1. The summary energy of vacuum is zero, but energy fluctuations of vacuum produce the negative and positive eners. The eners got
the impulse from other eners and change directions and speeds. Ener diameter is very small in comparison of space volume, the probability of perfect hit one to another is small (small cross section area). The eners become enough stable and the existing as real particles. If outer impulse p was small, the pair A, B produces the neutral pairs of real particles (fig. 2d). If outer impulse p is small, the pair A, B produces the neutral pairs of real eners like binary stars (fig. 2d). If outer impulse p is big, the particles A, B become the free the charged real eners (fig. 2e).

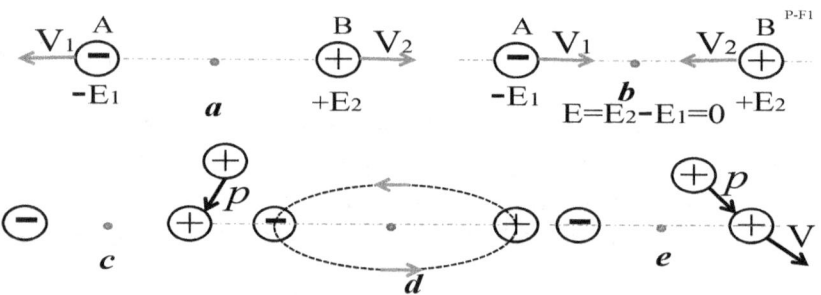

Fig.1. Virtual pairs of particles (eners A, B or "+" and "-") and their initial interaction: eners AA or BB - repulsion, AB or BA – attractive. *Notations:* a) Variations energy into vacuum (osculation the positive and negative energy); b) Annihilation (elimination) of pairs of opposed eners; c) One ener in pair gets the small side impact (impulse). d) Pairs of stable rotate relative to each other (like binary stars). e) One ener in pair gets a strong impact (impulse). The ener leaves the couple.

There is balance between relative stable and annihilate eners in vacuum. That balance can depend upon fields in vacuum.

Eners and Universe.

The offered eners model explains the emergence and expansion of the Universe. Before beginning of our Universe we did not have **nothing**: no space, no time, no any particles, no Universe. The appearance the couple (A – B) **eners** was beginning the produsion the space and time that is Universe. We don't know how this initial pair **eners** is appeared.

Our model the appearance of our Universe is principal difference from appearance of the current used model of Universe. In current model the University was created by Big Bang. The **Big Bang** theory is the prevailing cosmological model for the universe from the earliest known periods through its subsequent large-scale evolution. The model accounts for the fact that the universe expanded from a very high density and high temperature state, If the known laws of physics are extrapolated beyond where they are valid, there is a singularity. Extrapolation of the expansion of the universe backwards in time using general relativity yields an infinite density and temperature at a finite time in the past. The universe today is dominated by a mysterious form of energy known as dark energy, which apparently permeates all of space. The observations suggest 73% of the total energy density of today's universe is in this form.

The Big Bang theory depends on two major assumptions: the universality of physical laws and the cosmological principle. The cosmological principle states that on large scales the universe is homogeneous and isotropic. These ideas were initially taken as postulates.

As with any theory, a number of mysteries and problems have arisen as a result of the development of the Big Bang theory. Some of these mysteries and problems have been resolved while others are still outstanding.

Main problem of Big Bang is initial singularity: the entire universe at one point (??!!), super-gigantic energy, density, temperature? Where could it appear?

According to the known limitations of the applicability of modern physical theories, the earliest moment, enables the description, is the moment of the Planck epoch a temperature of about 10^{32} K (Planck temperature) and a density of about 10^{93} g / cm^3 (Planck density). The early universe was a highly homogeneous and isotropic medium with an extremely high energy density, temperature and pressure. We can not modelling what was before early Plank time 10^{-43} sec.

The offered model of the origin of Universe does not have this problem. One not requires energy (matter), not require anything for creating of Universe, except one pairs of very small eners. They produce initial space and time. The initial space and time produced (and continue to produce) our current Universe and the cycle of birth and destruction of eners within it it. The eners generated and is generating now matter and interactions.

Mass, charge and fundamental interactions.

The most known particles and their field, interaction may be created from ener compositions A, B of different forms, structures, mixture and density. Some possible construction from eners are sown in fig.2. Constructions (structures) from eners are: a) Neutral couple; b) Charged triple; c) Thread (rod) from eners; d) Magnetic construction; e) Complex charged construction; f) Plate charged construction; g) Volume construction; h) Circle construction; i) Spheric construction and shell (cover) construction.

Structures can be stable and unstable. We are interested in only stable structures having the lifetime of hundreds of seconds and more.

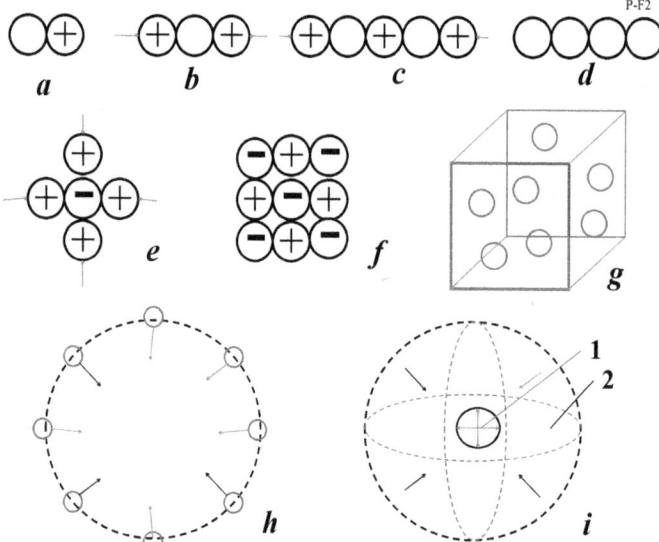

Fig.2. Constriction (structures) from eners. *Notations:* a) Neutral pairs of; b) Charged triple; c) Tread (rod) from eners ; d) Magnetic construction; e) Complex charged constriction ; f) Plate charged construction; g) Volume construction; h) Circle construction; i) Spherical construction (1 – core, 2 – pressure cloud from eners).

Let us show it for two important particles: mass particle and charge particle.

Mass.

The vacuum is produces very small eners (A and B) different signs "+" and "-", which attract one to other. Most of them have different directions and speeds, stable and cannot eliminate each other because neighboring eners decrease attraction and the Heisenberg uncertainty and Paulie exclusion principles prevent annihilation at close distance.

Let us take the structure (eners core or cloud in spherical form). Any structures have a small excess of positive or negative eners. That means they will be attractive one to other, condense, collect in bigger structures. The ener cloud will collapse and press to inner core. We get mass particles with high binding energy and negative gravitation field. This process is same on a larger scale getting the stars and planets from cosmic gas and dust into the Universe. The excess or shortage is very small random values. That way the gravitation force is small in comparison with charge force.

Mass and its properties appear when in small volume the eners concentration (energy density) reaches the density about $\rho_e = E/m = c^2$ (where ρ_e is density of the **binding** energy, E is **binding** energy, m is mass, c is light speed).

This density of energy creates the potential mass (ener) field from eners which scientists named gravitation field. That is scalar field connected to a mass particle which moves **together** with the mass particle (or collection of mass particles). The light sent from moving mass moves **into** gravitation field of this mass and has speed c of light. The well-known experiment of Michelson-Morlry confirms it (constant the light speed in moving inertial system). Moreover the Special Theory of Relative is built on it. Creation of mass particles from A, B eners is a stochastic process. That means that mass particles from them have the small difference in amounts of the eners A, B and attracting one to the other. This is gravity interaction. The gravity interaction of mass particles is small in comparison with other interaction. But a large mass produces a large gravitation force.

The offered model of mass interaction easily explains the old scientific discussion: What is the distinction between gravitation, inertial, centrifugal masses? If the mass moves uniformly in a straight line (inertial system) the gravity field is moving together with mass and no braking of mass (fig. 3a). If mass is accelerated, the observer in former system shows the thickening of living (scalar) lines (density of gravity field increases in ahead body and decreases in back of body) and brake force appears (fig. 3b). The same situation if mass moves in circle line (fig. 3c). In this case, the density of the mass field inside circle increase, out of circle - decrease and the centrifugal force appears (fig. 3c).

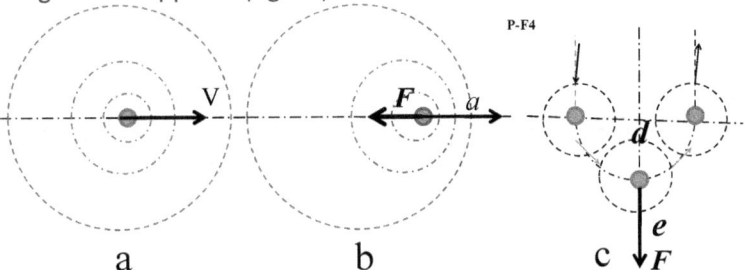

Fig.3. a) Moving mass with constant speed *V*. The ener field created by planet is moved together with planet, having equal density, force and light speed do not change; b) The body has acceleration *a*. The density of field ahead of body increases and appear a braking force *F*; c) When the body is moving in a circle the field density inside of circle increases, the outside decreases and appear as centrifugal force *F*.

The gravitation negative potential field of two planets between them is sum of two negative values. The sum is less than the initial value. That way the density of the summary gravitation field is less between planets and they attract one to other.

Charge.

The special construction from eners A-B creates the charge. If construction has much more A then B, we have one charge (for example, "+"), if conversely – we have other charge ("-"). The scalar field of opposed charges have an opposed signs. The charges produce the special charge fields from eners, which interact ONLY with charge fields of other charges. The two same charges fields increase the sum field density of between same charges and repel same charges. The opposed charge decrease the charge density between charges and attract the opposed charges. Summary: the same charges (++ or - -) repel each other. The different charges attract each other.

In charge constructions (structures) the same charged particles are kept together in core 1 the shell (cloud) from eners 2 (fig. 4a). But the requested pressure of shell is much more then pressure for keeping the mass structures. That way no the stable big charge structures having a big excess the same charges.

The charge field as mass field has long distance interaction and potential field.

The main binding energy of mass construction (mass + field) is concentrated in mass, the main energy of charge is concentrated in the charge field.

Creating the quarks from eners.

If we created the charge construction +1/3e, -1/3e from A and B ("-" and "+") eners as it is described over, it is no problem to create the quarks from them. For example, let us marked the charged construction having -1/3e ener charge as C_-, the charged construction having +1/3e as C_+. The quark u_{up} having the charge +2/3e may be created from two C_+ ($C_+ C_+$) plus the shell (cover) from the eners. The summary binding energy is about 2.3 MeV/c^2. The quark d_{down} having the charge -1/3e may be created from one C_- (C_-) plus the shell (cover) from the eners. The summary binding energy is about 4.8 MeV/c^2.

As known quarks may be created from proton (**uud**) and neutron (**udd**). Quarks are keeping together by the strong shell from eners collected from eners vacuum. That way the summary energy of proton (938 MeV) and neutron (939 MeV) is much more than energy of three quarks (9.4 MeV, 11.9 MeV). The very strong shell around the quark collections explain the confinement – inability to get quarks from nucleons.

Similarly, we can design main stable particles: charged electron ($C_- C_- C_-$) and neutral neutrino (A+B, fig.1c). The antiparticles can be got the changing the all eners A to B and B to A. They have a same property.

In annihilation, the most blind energy of eners in construction will be converted at radiation (photon). The radiations (photons) are oscillation of the eners field in vacuum.

Most other unstable particles may be designed from the eners A, B. The eners are generated the gravitation and electric fields. They transfer the radiation.

Strong and weak nuclear interaction.

The strong and weak nuclear interaction can be explained by pressure ener shell around nucleons. As it is shown above the cover from attractive eners collapse and has pressure to center and surface tension. The one spherical shall has less surface than two shells having same (or less) sum volume. The nuclear shall of simple nuclear core when they close one to other can fuse together (fig.4b) because the volume and energy one sphere is less than two. As result we have a fuse nuclear reaction. If nuclear core has a complex construction, contains a lot of nucleus and positrons or small shell, the ener shell cannot to keep intact particles, so begins nuclear decay and instability (fig. 4c).

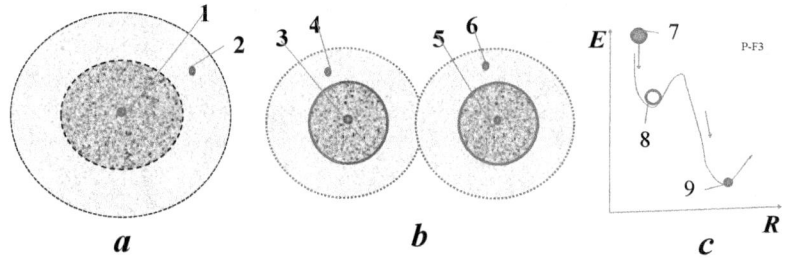

Fig.4. Strong and weak nuclear forces (interactions). *Notations*: a) Construction (structure) 1 is covered by shell (cloud) 2 from the emers (A, B). The cloud 2 are pressing the structure 1 and create its stability. b) If two shell get a contact they may merge two structures in one, because energy of one structure may be less than energy of two structures; c) Possible stable summary potential energy of construction from distance between them. 1, 3, 5 are internal structures; 2, 4, 6 are shells; 7 is structure; 8 is local minimum of energy; 9 is global minimum of energy. Stable (lifetime) of particles repents from hole depth in potential field and fluctuation energy of particles.

Features of the proposed preon model.

The offered preon model has next features:
1. Only two particles (eners: A, B) having opposed energy.
2. Only two interactions (attraction between different eners and repulsion from same eners).
3. All short distance interactions may be designed from eners (as pressure shell, cover of eners construction).
4. The mass and positive-negative charges (and their long distance field) may be design from eners.
5. All known quarks, leptons and neutrino may be designed from eners. As known, the nucleons are designed from quarks.
6. All main stable particles and most unstable particles may be design from eners.
7. In long distance interaction ($1/r^2$) of electric and gravity forces interact (add and subtract intensities) the scalar electric and gravity fields, not use photons and gravitons.

Some Results:

From suggested preon model follows:
1. All three types of mass (gravity, acceleration-centrifugal, passive) are same.
2. All matters are designed from two eners.
3. All four main interactions (gravity, electric, weak and strong nuclear) are designed from two eners.
4. It is impossible to design the negative mass.
5. It is impossible to create the anti-gravitation.

6. Simultaneous replacement of the particles A to B and B to A does not change the Universe.
7. Antiparticles connect with Law of Symmetry (change A by B).
8. From point 7 follows:
  a) The B ener is antiparticle of A ener;
  b) The mass particle is same as its anti-mass particle;
  c) The negative electric charge particle is antiparticle of the positive electric charge particle (example: electron and positron).

Difference between eners and rishon

Our preon model is different from the rishon model. There are substantial difference between rishons and eners. Eners two (not three) and eners have interaction one to another. They have energy. Rishon can be designed from eners, not the reverse. Eners theory explain an appearance mass, charges and all interactions between particles (gravitation, electromagnetic, strong and week nuclear and possible future interaction), the reshon theory cannot do it. Eners theory explain many scientific facts and phenomena. And so on.

Rishon theory simply divides quarks into its component parts and states that they are made up of these parts.

AB Preon Interaction Theory and current theory of interaction.

Current theory of interaction assumes the curriers of interaction (gravitation, strong and weak interaction, electro-magnetic). They are bosons-particles (gravitons, gluons, W, Z bosons, photons and Higge boson). They are running between bodies and pass forces from one body to other. That model has a lot of questions. Gluon and photon do not have mass and cannot transfer the momentum (force). Photon has constant speed c and also cannot to pass momentum. Moreover, this method can transfer only the repulsion force, not attraction.

AB theory assumes between bodies there are specific long distance scalar fields (gravity, charge) created these bodies. Those fields have a different density and interact with specific field and bodies. Increasing (summation) of density produces the repulsion, decreasing (subtraction) of density produces attraction.

Approximately long-range field looks like a field between two charged particles (fig. 5). The lines between same charged particles condense (fig. 5b). The lines between opposed charged particles elongate (fig. 5a). Result: different charged particles (eners) attract, same charged particles - repel. The field is transferring this situation (forces) to bodies. The gravitons (not open yet) and photons are only the environment of the special bodies (mass, charge).

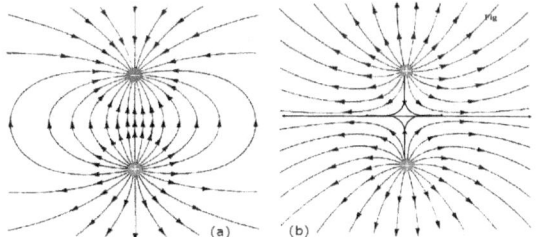

Fig.5. The interaction between two charged particles. a) same charged; b) different charged.

The mechanism of strong and weak interactions is another. In strong interaction, the power shells covering the assembly of quarks – nucleons . In special cases the shell can merge (connect, unit) and keep together the different nucleus. If nucleus are light and have a few particles, they excrete a binding energy. If there are a lot of particles in nucleus, the eners shell cannot keep them together and nucleus decays. If summary blind energy of new parts is less, the surplus is allocated.

The weak interaction works the same.

Theory and useful equations.

Below author gives the information which are useful for further developing the offered preon model.

The main fields are acceleration, gravity, electric, magnetic and photon/radiation. Density of energy in given point of these fields compute by equations [5]:

$$w_a = \frac{1}{G}\frac{a^2}{2}, \quad w_g = \frac{1}{G}\frac{g^2}{2}, \quad w_e = \varepsilon_0 \frac{E^2}{2}, \quad w_m = \mu_0 \frac{H^2}{2}, \quad w_r = \frac{\sigma}{c}t^4, \quad w_E = \frac{c^2}{GT^2}, \quad (1)$$

where w_a is density of acceleration energy, J/m³; w_g is density of gravitation energy, J/m³; w_e is density of electric energy, J/m³; w_m is density of magnetic energy, J/m³; w_r is density of radiation energy, J/m³; w_E is time energy density, J/m³. a is acceleration, m/s²; g is gravitation, m/s²; $\sigma = 5.67 \cdot 10^{-5}$, W/m²K is Stefan – Boltzmann constant, W/m²K ; E is electric intensity, V/m or N/C; H is magnetic intensity, T or Vs/m² or Wb/m²; w_r is density of radiation energy, J/m³; t is temperature , K; T is time, sec. The last two formulas show the energy density depends from temperature and time. $\mu_0 = 1.257 \cdot 10^{-6}$, H/m; $\varepsilon_0 = 8.854 \cdot 10^{-12}$, F/m.

Full energy, W, we find by integration of density to a full volume.

$$W = \int_V w dv \quad (2)$$

These computations in analytical form we can take as relating to simple geometric figures as, for example, the spherical forms of fields.

Binding energy of spherical mass equal density includes three components:
Field energy of internal part of sphere i

$$W_{g,1} = \frac{\pi G M^2}{5R}, \quad (3)$$

Here $W_{g,1}$ field energy of internal part of sphere, J; $G = 6.672 \cdot 10^{-11}$ Nm²/kg² –gravity constant; M – mass, kg; R – radius sphere, m.

Field energy of excternal part of sphere is

$$W_{g,2} = \frac{2\pi G M^2}{R}, \quad (4)$$

Mass energy is

$$E = Mc^2, \quad (5)$$

where $c = 3 \cdot 10^8$ m/s is light speed.

The field energy of nucleon is on 22 orders less then E. For Earth it is less on 9 orders, for Sun it is less on 3 orders and for Black Hole it is in π times more then an energy mass of the Black Hole.

For electric charge the outer energy of electric field significantly depends from unknown radius R of charge. One can be computed by equation

$$W_{e,2} = k\frac{Q^2}{2R}, \quad (6)$$

where $k = 9 \cdot 10^9$ is electric constant, Nm²/C²; Q is electric charge, C. The distance of gravity and electrostatic interaction is infinity. But force decreases as $1/r^2$. Classical radous of electron is $R = 2.8179 \cdot 10^{-15}$ m, charge $e = 1.6 \cdot 10^{-19}$ C.

The Energy of the strong interaction may be computed by Yukawa equation.
The Yukawa potential (also called a screened Coulomb potential) is a potential of the form

$$V_{\text{Yukawa}}(r) = -g^2 \frac{e^{-\mu r}}{r}, \quad (7)$$

where g is a magnitude scaling constant, i.e., the amplitude of potential, μ is the Yukawa particle mass, r is the radial distance to the particle. The potential is monotone increasing, implying that the force is always attractive. The constants are determined empirically. The Yukawa potential depends only on the distance between particles, *r*, hence it models a central force.

The interaction distance is only 10^{-15} m. Fig.6 shows the potential energy (MeV) and force (10^4 N) between nucleons.

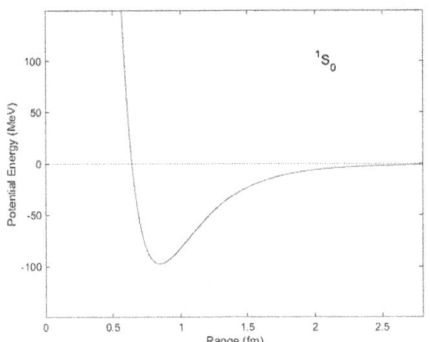
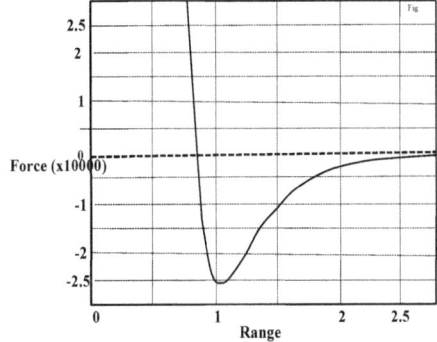

Fig.6. Potential energy and strong force of nucleon via distance from center of nucleon.
The negative energy is attraction, the positive energy is repulsion.

For weak interaction potential is closed to strong interaction (7)

$$V = C \frac{e^{-m_{W,Z} r}}{r}, \qquad (8)$$

Where $m_{W,Z}$ is mass of gauge particle W, Z; *C* is constant. But interaction distance is very small 10^{-18} m.

The energy of photon is

$$E = h\nu, \qquad (9)$$

where $\hbar = 6.626 \cdot 10^{-34}$, J·s is Plank constant, ν is frequency, 1/s. Radio frequency has $\nu = 3 \cdot 10^8$; X-ray has frequency up $3 \cdot 10^{27}$.

Density of radiation energy is

$$w = \frac{\varepsilon_o E^2}{2} + \frac{\mu_o H^2}{2}. \qquad (10)$$

Specific pressure of radiation is $p = w$. Active distance is infinity. Radiation energy, density and pressure decreases as $1/r^2$.

Note: The construction from eners (body) creates around self the energy field and radiation.

Discussion

Criticism of the current models of Universe and Micro-World. In current time the most scientists believe: the Universe appeared in Big Bang and Inflation. Initial World was a singular point (infinity small volume of dot, infinity density of energy and matter). This point was exploded about 14 billions years ago, rapidly expanded (speed significantly more *c*) and created stars, galactic, radiation, cosmic dust, etc. The proof is radiation 2.7K of Big Bang in space. This model has a lot of questions not having answers: How did infinity point appear amid gigantic energy and matter? Why did it explode? Where initial point was located? Why radiation having maximal known speed *c* not left our place during 14 billion years? Why the universe is expanding in present time? And so on.

Interaction in **Micro-World** the current theory explains the following way: world consists of particles. When one particle knows (how? Radio-location?) about other particle, one gives birth (how?) to a special carrier particle (who is designer?), give it momentum (how? Where does it take energy?) and sends it to other particles. The other particle accepts momentum; gives birth to same carrier particles and send it back [1] p.91. For example, in weak interaction the carrier particles are W$^+$, W$^-$, Z^0 gauge bosones. They have mass 80-90 GeV. It is in 100 times more then mass of nucleus (939 MeV)! Higgs boson has 125 GeV. From where does the nucleus take this high energy? Carrier of strong interaction glue (and photon) has a zero mass. How can it pass impulse (momentum)? Moreover, these carrier-particles can only transmit a repulsive force. And what about the force of attraction?

The offered pion AB model allows explaining from one position all current and future interactions.

Brief description of the suggested model. Author used the Principle of Parsimony or Simplicity. This Principle is: *Simplicity is base of Universe*. The Local Principle Simplicity is: *Any physical phenomenon is simplicity in particular area.*

Author offers the preon model of microworld having one pair initial virtual particles - eners (A,B or "+","-") and one pair main interaction (attraction and repulsion) between them. The pairs of A-B ener has opposed the equal deviations of zero energy and momentum. They have short life - swiftly annihilate. But due to the impact of other eners and random factors the eners become stable.

The pairs of A-B has volume. They birth the same pair A-B and so on – produce space filled with the eners. The time is local speed of the interaction between eners (maximal transferring of information). The eners have possibility to create the stable constructions. Some of them under certain conditions get the special properties and create an own field which interacts with the same or related structures. For example, when the ener construction reaches the density of a blind energy c^2, they get a mass and the gravitation field. When density of negative or positive eners has certain density (charged blind energy), they get a charge and electric field.

As the author shown, the eners allow the design of the main stable quarks and interaction between them. It means we can design all hadrons and leptons – the main stable matters and main interactions.

The offered model has big possibilities to create the known and future particles and their interaction. I call the scientists to develop this model. The eners may be the revealed face of dark energy and dark matter.

Brief Results.

Author using the Principle of Parsimony (Occam's razor) offers ideas for creating a simple preon model of the University and Micro-World. That is only pair of fundamental particles named ***eners*** (A-B, or "+". "-") and pair the fundamental interaction (attraction, repulsion).

He shortly shows how we can design from eners: mass, charge, quaks, matter, gravitation and charge (electric) fields, and strong-weak nuclear interactions. He shows how the Universe can be designed from eners without Big Bang. Why the universe is expanding. The eners may be good candidates for dark energy and dark matter.   Some results are following from the suggest model to support current experiments or may be checked up by future experiments.

- All three types of mass (gravity, acceleration, centrifugal) are same.
- It is impossible to design the negative mass.
- It is impossible to create the anti-gravitation.
- Antiparticles connect with Law of Symmetry (change A by B).
- The mass particle is same its anti-mass particle;
- The negative electric charge particle is antiparticle of the positive electric
  charge particle (electron and positron).

References

1. Хокинг С., Краткая история времени, Санк-Петербург, АМФОРА, 2015, стр. 91. Translation to Russia from Hawking S., A Brief History of Time.
2. Harari, H. (1979). "A Schematic Model of Quarks and Leptons" (PDF). Physics Letters B **86** (1): 83–86. Bibcode:1979PhLB...86...83H. doi: 10.1016/0370-2693(79)90626-9.
3. Shupe, M. A. (1979). "A Composite Model of Leptons and Quarks". Physics Letters B **86** (1): 87–92. Bibcode:1979PhLB...86...87S. doi: 10.1016/0370-2693(79)90627-0.
4. Zenczykowski, P. (2008). "The Harari–Shupe preon model and nonrelativistic quantum phase space". Physics Letters B **660** (5): 567–572. arXiv:0803.0223. Bibcode:2008PhLB..660..567Z. doi:10.1016/j.physletb.2008.01.045.
5. Bolonkin A.A., Universe (Part 3). Relations between Charge, Time, Matter, Volume, Distance, and Energy. 30/12/2013. The General Science Journal, #5245. , http://viXra.org/abs/1401.0075 .

25 February 2016.

Chapter 2

Universe (Part 3). Relations between Time, Matter, Space, Charge, and Energy

Summary

In Universe (Part 1)[1] author has developed a theory which allows derivation of the unknown relations between the main parameters (energy, time, volume, matter) in the Universe. In given part 3 he added charge as main parameter in this theory. He finds also the quantum (minimal values) of energy, time, volume and matter and he applied these quantum for estimations of quantum volatility and the estimation of some values of our Universe and received both well-known and new unknown relations.

Author offers possibly valid relations between charge, time, matter, volume, distance, and energy. The net picture derived is that in the Universe exists ONLY one substance – ENERGY. Charge, time, matter, volume, fields are evidence of this energy and they can be transformed one to other. Author gives the equations which allow to calculate these transformation like the famous formula $E = mc^2$. Some assumptions about the structure of the Universe follow from these relations.

Most offered equations give results close to approximately known data of Universe, the others allow checking up by experiment.

Key words: Universe, time, matter, volume, distance, energy; limits of specific density of energy, matter, pressure, temperature, intensity of fields; collapse of space and time into point.

Introduction

In the theoretical physic the next fundamental constants presented in Table 1 are important.

Table 1: Fundamental physical constants

Constant	Symbol	Dimension	Value in SI units with uncertainties
Speed of light in vacuum	c	$L\,T^{-1}$	2.99792458×10^8 m s^{-1}
Gravitational constant	G	$L^3\,M^{-1}\,T^{-2}$	$6.67384(80) \times 10^{-11}$ m^3 kg^{-1} s^{-2}
Reduced Planck constant	$\hbar = h/2\pi$ where h is Planck constant $h = 6.625\,068\,76(52) \times 10^{-34}$	$L^2\,M\,T^{-1}$	$1.054571726(47) \times 10^{-34}$ J s
Coulomb constant	$(4\pi\varepsilon_0)^{-1}$ where ε_0 is the permittivity of free space $\varepsilon_0 = 8.854\,187\,817... \times 10^{-12}$	$L^3\,M\,T^{-2}\,Q^{-2}$	$8.9875517873681764 \times 10^9$ kg m^3 s^{-2} C^{-2} *(exact by definitions of ampere and meter)*
Boltzmann constant	k_B	$L^2\,M\,T^{-2}\,\Theta^{-1}$	$1.3806488(13) \times 10^{-23}$ J/K

Where are: L = length, M = mass, T = time, Q = electric charge, Θ = temperature.

If we take these constants as base units, we get the Planks units:

Table 2: Base Planck units

Name	Dimension	Expression	Value (SI units)

Planck length	Length (L)	$l_P = \sqrt{\dfrac{\hbar G}{c^3}}$	$1.616\,199(97) \times 10^{-35}$ m
Planck mass	Mass (M)	$m_P = \sqrt{\dfrac{\hbar c}{G}}$	$2.176\,51(13) \times 10^{-8}$ kg
Planck time	Time (T)	$t_P = \dfrac{l_P}{c} = \dfrac{\hbar}{m_P c^2} = \sqrt{\dfrac{\hbar G}{c^5}}$	$5.391\,06(32) \times 10^{-44}$ s
Planck Charge	Electric charge (Q)	$q_P = \sqrt{4\pi\varepsilon_0 \hbar c}$	$1.875\,545\,956(41) \times 10^{-18}$ C
Planck temperature	Temperature (Θ)	$T_P = \dfrac{m_P c^2}{k_B} = \sqrt{\dfrac{\hbar c^5}{G k_B^2}}$	$1.416\,833(85) \times 10^{32}$ K

From data Table 2 we can receive the derived Planck units (Table 3).

Table 3: Derived Planck units

Name	Dimension	Expression	Approximate SI equivalent
Planck area	Area (L^2)	$l_P^2 = \dfrac{\hbar G}{c^3}$	2.61223×10^{-70} m^2
Planck volume	Volume (L^3)	$l_P^3 = \left(\dfrac{\hbar G}{c^3}\right)^{\frac{3}{2}} = \sqrt{\dfrac{(\hbar G)^3}{c^9}}$	4.22419×10^{-105} m^3
Planck momentum	Momentum (LMT^{-1})	$m_P c = \dfrac{\hbar}{l_P} = \sqrt{\dfrac{\hbar c^3}{G}}$	6.52485 kg m/s
Planck energy	Energy (L^{2}MT^{-2})	$E_P = m_P c^2 = \dfrac{\hbar}{t_P} = \sqrt{\dfrac{\hbar c^5}{G}}$	1.9561×10^9 J
Plank force	Force (LMT^{-2})	$F_P = \dfrac{E_P}{l_P} = \dfrac{\hbar}{l_P t_P} = \dfrac{c^4}{G}$	1.21027×10^{44} N
Planck power	Power (L^{2}MT^{-3})	$P_P = \dfrac{E_P}{t_P} = \dfrac{\hbar}{t_P^2} = \dfrac{c^5}{G}$	3.62831×10^{52} W
Planck density	Density (L^{-3}M)	$\rho_P = \dfrac{m_P}{l_P^3} = \dfrac{\hbar t_P}{l_P^5} = \dfrac{c^5}{\hbar G^2}$	5.15500×10^{96} kg/m^3
Planck energy density	Energy density (L^{-1}MT^{-2})	$\rho_P^E = \dfrac{E_P}{l_P^3} = \dfrac{c^7}{\hbar G^2}$	4.63298×10^{113} J/m^3
Planck intensity	Intensity (MT^{-3})	$I_P = \rho_P^E c = \dfrac{P_P}{l_P^2} = \dfrac{c^8}{\hbar G^2}$	1.38893×10^{122} W/m^2
Planck angular frequency	Frequency (T^{-1})	$\omega_P = \dfrac{1}{t_P} = \sqrt{\dfrac{c^5}{\hbar G}}$	1.85487×10^{43} s^{-1}

Planck pressure	Pressure ($L^{-1}MT^{-2}$)	$p_P = \dfrac{F_P}{l_P^2} = \dfrac{\hbar}{l_P^3 t_P} = \dfrac{c^7}{\hbar G^2}$	4.63309×10^{113} Pa
Planck currency	Electrictric currency (QT^{-1})	$I_P = \dfrac{q_P}{t_P} = \sqrt{\dfrac{4\pi\epsilon_0 c^6}{G}}$	3.4789×10^{25} A
Planck voltage	Voltage ($L^2MT^{-2}Q^{-1}$)	$V_P = \dfrac{E_P}{q_P} = \dfrac{\hbar}{t_P q_P} = \sqrt{\dfrac{c^4}{4\pi\epsilon_0 G}}$	1.04295×10^{27} V
Planck impedance	Resistance ($L^2MT^{-1}Q^{-2}$)	$Z_P = \dfrac{V_P}{I_P} = \dfrac{\hbar}{q_P^2} = \dfrac{1}{4\pi\epsilon_0 c} = \dfrac{Z_0}{4\pi}$	29.9792458 Ω

Universal units do not depend from Earth units. That is suitable for the Universe communication. They also give the more simple physical equations.

Theory. Relation between charge, time, matter, volume, distance and energy.

The author presents an original theory which allows derivation of unknown relations between main parameters in a given field of nature. He applies his hypotheses to theory of Universe. The next well-known constants used in his equations are below:

$c = 2.997925 \cdot 10^8 \ m/s$, $e = 1.60219 \cdot 10^{-19} \ C$, $G = 6.6743 \cdot 10^{-11} \ m^3/kg \cdot s^2$,

$\varepsilon_0 = \dfrac{1}{36\pi \cdot 10^9} = 8.8542 \cdot 10^{-12} \ \dfrac{F}{m}$, $k = \dfrac{1}{4\pi\varepsilon_0} = 8.987551787 \cdot 10^9 \ \dfrac{kg \cdot m^3}{s^2}$,

$\mu_0 = 4\pi \cdot 10^{-7} = 1.257 \cdot 10^{-6} \ \dfrac{H}{m}$, $h = 6.6261 \cdot 10^{-34} \ \dfrac{kg \cdot m^2}{s}$, $\hbar = h/2\pi$,

$\sigma = 5.67032 \cdot 10^{-8} \ W/m^2 K$, $\pi = 3.141592654$, $k_B = 1{,}3896503(24) \cdot 10^{-23} \ J \cdot K^{-1}$ (1)

where c is speed of light in vacuum, m/s; e is electronic charge, C; G is a Newton gravitation constant, Nm²/kg²; ε_0 is electric constant, F/m; μ_0 is magnetic constant, H/m; h is Planck constant, J·s; σ is Stefan – Boltzmann constant, W/m²K ; k_B is Boltzman constant, J/K.

In our equations below the constant G is constant of gravitation theory, c is constant of relativistic theory, and h is constant of quantum theory. Productions $Ghc^5 \approx 1.071$ and $kc^4 \approx 1.088 \times 10^{52} G$ produce the new constants:

$$Ghc^5 \approx 1, \quad kc^4 \approx 10^{52} G, \qquad (2)$$

which we will use in our equation.

The author postulated the following relations:
1. Relations between time, matter, volume, distance, specific density of matter and energy :

$$T = \dfrac{G}{c^5} E, \quad T = \dfrac{G}{c^3} M, \quad T = c^{-1} v^{1/3}, \quad T = \dfrac{R}{c}, \quad T = \left(\dfrac{kG}{c^2}\right)^{1/2} Q, \quad T = G^{-1/2} \rho^{-1/2},$$

or $T = 2.756144 \cdot 10^{-53} E$, $T = 2.47709939 \cdot 10^{-36} M$, $T = 2.583467 \cdot 10^{-9} Q$,

$T = 3.33564 \cdot 10^{-9} R$, $T = 2.2448563 \cdot 10^{-24} \rho^{-2}$, (3)

where T is time in sec; E is energy in J; M is mass, kg; v is volume in m³; R is distance, m; ρ is specific density of matter in given volume, kg/m³, Q is charge, C. (Only the first 6 digits are right in all our formulas).

The dimensional theory is employed; that way these relations are obtained to within a constant factor. That factor may be derived from experiment. This factor has been neglected in cosmology and high energy physics. But these equations (2)-(6) cannot be derived ONLY from dimensional theory because dimensional theory does not contain the physical constants.

Equations (3) may be written in form

$$E = \frac{c^5}{G}T, \quad M = \frac{c^3}{G}T, \quad v = c^3 T^3, \quad R = cT, \quad Q = \left(\frac{c^2}{kG}\right)^{1/2} T, \quad \rho = 1/(GT^2),$$

or $E = 3.62825745 \cdot 10^{52} T, \quad M = 4.454628 \cdot 10^{35} T, \quad \rho = 1.5 \cdot 10^{10}/T^2.$ (4)

From these equations follow some interesting propositions. Time is energy, Time depends upon mass, volume, length, electric charges and density of matter. If time simultaneously produced the positive and negative charges, the total charge is zero. Time can create the energy, mass, distance, volume change and the density of matter in the Universe or the energy produce time, matter, distance, volume and charge (positive and negative simultaneously)(see (9)-(10)).

If we will use the relation (2) (quantum constant) the relation between time and the energy, mass, distance, volume and change, may be written in form

$$T \approx hG^2 E, \quad T \approx hG^2 c^2 M, \quad T^3 \approx hGc^2 v, \quad T \approx hGc^4 R, \quad T \approx \left(\frac{khG^2}{c}\right)^{1/2} Q,$$ (5)

2. Relations between volumes, energy, matter, time, and distance

$$v = \frac{G^3}{c^{12}} E^3 \quad v_n = \frac{G^n}{c^{4n}} E^n, \quad v = c^3 T^3, \quad v = \frac{G^3}{c^6} M^3, \quad v = \frac{4\pi}{3} R^3,$$

or $v = 5.64115466 \times 10^{-133} E^3, \quad v = 2.694401 \times 10^{25} T^3, \quad v = 4.095365 \times 10^{-82} M^3,$ (6)

where v is volume of 3-demantional space, m³; v_n is n-dimensional space, mⁿ.

3. Relations between matter, time, volume, distance, energy, charge and temperature are

$$M = \frac{c^3}{G}T, \quad M = \frac{c^2}{G}v^{1/3}, \quad M = \frac{c^2}{G}R, \quad M = \frac{1}{c^2}E, \quad M = \left(\frac{k}{G}\right)^{1/2} Q, \quad M = \left(\frac{k_B}{G}\right)^{1/2} t,$$

$$M = 4.0369797 \times 10^{35} T, \quad M = 1.34659 \times 10^{27} v^{1/3}, \quad M = 1.34659 \times 10^{27} R,$$

$$M = 1.40895 \times 10^{20} Q, \quad M = 2.068058 \times 10^{-13} t.$$ (7)

where t is temperature, K; k_B is Boltzmann constant, J/K.

If we will use the relation (2) (quantum constant) the relation between mass, time and the energy, distance, volume and change, may be written in form

$$M \approx \frac{1}{hG^2 c^2} T, \quad M \approx \frac{1}{hG^2 c^3} R, \quad M \approx hGc^3 E,$$

$$M \approx (khc^5) Q, \quad M = k_B hGc^3 t,$$ (8)

4. We can receive from equations (3) - (8) the expressions for the energy from time, volume, distance, mass and charge

$$E = \frac{c^5}{G}T, \quad E = \frac{c^4}{G}v^{1/3}, \quad E_n = \frac{c^{4n}}{G^n} v_n^{1/n}, \quad E = \frac{c^4}{G}R,$$

$$E = k_B t, \quad E = \left(\frac{kc^4}{G}\right)^{1/2} Q, \quad E = c^2 M,$$

$$E = 3.62825745 \cdot 10^{52} T, \quad E = 1.2102562 \cdot 10^{44} v^{1/3},$$

$$E = 1.2102562 \cdot 10^{44} R, \quad E = 8.98755 \cdot 10^{16} M, \quad E = 1.0429 \cdot 10^{26} Q.$$ (9)

Here t is temperature, K.

Last equation in (9) is the well known relation between energy and matter. This relationship follows from (2) – (8) as a special case. This indirectly confirms the correctness of the equations (2) – (9) as a special case.

If we will use the relation (2) (quantum constant) the relation between energy, mass, distance, volume, change and temperature may be written in form

$$E \approx \frac{1}{hG^2}T, \quad E \approx \frac{1}{hG^2c}v^{1/3}, \quad E \approx \frac{1}{hG^2c}R, \quad E = Mc^2, \quad E \approx \left(\frac{k}{hG^2c}\right)^{1/2}Q, \qquad (10)$$

Here v (Latin) is volume, m³.

5. The relations between the density of matter, energy, charge and time (frequency) are following:

$$\rho_M = \frac{1}{G}\frac{1}{T^2}, \quad \rho_M = \frac{1}{G}v^2, \quad \rho_E = \frac{h}{c^3}\frac{1}{T^2}, \quad \rho_E = \frac{h}{c^3}v^2, \quad \rho_Q = \left(\frac{h}{kc^5}\right)^{1/2}T^2, \quad \rho_Q = \left(\frac{h}{kc^5}\right)^{1/2}\frac{1}{v^2}, \qquad (11)$$

Where ρ_M, ρ_E, ρ_Q are density of matter, energy and charge respectively, kg/m³, J/m³, C/m³; v (Greg) is frequency, 1/s.

If we will use the relation (2) (quantum constant) the relation between the density of the matter, energy, and change, may be written in form:

$$\rho_M \approx \frac{1}{hc^5}\frac{1}{T^2}, \quad \rho_M \approx \frac{1}{hc^5}v^2, \quad \rho_E \approx Gh^2c^2\frac{1}{T^2},$$

$$\rho_E \approx Gh^2c^2v^2, \quad \rho_Q \approx \left(\frac{Gh^2}{k}\right)^{1/2}T^2, \quad \rho_Q \approx \left(\frac{Gh^2}{k}\right)^{1/2}\frac{1}{v^2}, \qquad (12)$$

6. The relations between the field intensity of electric, magnetic and gravitation field a mnd time (frequency) are following:

$$E_e = \left(\frac{kh}{c^3}\right)^{1/2}T^2 = \left(\frac{kh}{c^3}\right)^{1/2}\frac{1}{v^2}, \quad H = \frac{\varepsilon_0}{c}E_e = \frac{c}{\mu_0}E_e, \quad H_G = (\varepsilon_0 G)^{1/2}E_e = \left(\frac{c^2G}{\mu_0}\right)^{1/2}E_e, \qquad (13)$$

Where E_e, H, H_G are intensity of electric, magnetic and gravitation fields respectively; ε_0, μ_0 are electric and magnetic constants (see (1)).

Application to current Universe

Let us estimate the real size and parameters (mass, radius, time, density, etc.) of the Universe. We can make it if we accurately know at least one of its parameters.

Thus the most reliable parameter is the lifetime of the Universe after the Big Bang. Estimates of the observed mass and radius are growing all the time. Estimation of the time specified is about 14 billion years now (13.75±0.17 billion years).

$$M = \frac{c^3}{G}T, \quad E = \frac{c^5}{G}T, \quad R = cT, \quad v = \frac{4}{3}\pi R^3, \quad \rho = \frac{1}{GT^2},$$

$$\text{or} \quad M = 4.0369787 \cdot 10^{35}T, \quad E = 3.62825745 \cdot 10^{52}T, \qquad (14)$$

$$R \approx 3 \cdot 10^8 T, \quad \rho = 1.5 \cdot 10^{10}/T^2.$$

Substitute in (14) the age of Universe after Big Bang (T=14 billions years = 4.4·10¹⁷ sec) we receive:

$$M = 1.78 \cdot 10^{53}\,kg > 1.4 \cdot 10^{53}\,kg, \quad E = 1.6 \cdot 10^{70}\,J,$$

$$R = 1.32 \cdot 10^{26}\,m < 4.4 \cdot 10^{26}\,m, \quad v = 10^{79}\,m^3, \quad \rho = 7.75 \cdot 10^{-26}\,kg/m^3 > 10^{-26}\,kg/m^3. \qquad (15)$$

In right side of the inequality (15) is given the estimations of universal parameters made by other researchers. They are very different. The author took average or approximate values.

As you see the values received by offered equations and other methods have similar magnitudes. The mass of the Universe is little more because we do not see the whole Universe (only the closer bodies). The estimation of radius is more than light can travel in the time since the origin of the Universe. It is possible the Universe in initial time had other physical laws than now or the expansion of space may account for this. The difference of space density is result of the old methods that do not include invisible matter, dark matter and dark energy.

The main fields are acceleration, gravity, electric, magnetic and photon/radiation. Density of energy in given point of these fields compute by equations:

$$w_a = \frac{1}{G}\frac{a^2}{2}, \quad w_g = \frac{1}{G}\frac{g^2}{2}, \quad w_e = \varepsilon_0 \frac{E^2}{2}, \quad w_m = \mu_0 \frac{H^2}{2}, \quad w_r = \frac{\sigma}{c} t^4, \quad w_E = \frac{c^2}{GT^2},$$ (16)

where w_a is density of acceleration energy, J/m³; w_g is density of gravitation energy, J/m³; w_e is density of electric energy, J/m³; w_m is density of magnetic energy, J/m³; w_r is density of radiation energy, J/m³; w_E is time energy density, J/m³. a is acceleration, m/s²; g is gravitation, m/s²; σ is Stefan – Boltzmann constant, W/m²K ; E is electric intensity, V/m or N/C; H is magnetic intensity, T or Vs/m² or Wb/m²; w_r is density of radiation energy, J/m³; t is temperature , K; T is time, sec. The last two formulas show the energy density depends from temperature and time.

Full energy, W, we find by integration of density to a full volume.

$$W = \int_V w dv$$ (17)

These computations in analytical form we can take as relating to simple geometric figures as, for example, the spherical forms of fields.

Note: In many cases the light speed "c" in the equations (2)-(13) may be changed in conventional speed V. That means we can verify the formulas (2)-(13) and find the correct constant factor.

Quanta of energy, charge, time, matter, volume, and distance.

It is known the energy of photon is

$$E_q = h\nu,$$

where ν is frequency, 1/s (ν = 1, 2, 3, …). The minimal quantum of photon energy is when ν = 1,

$$E_q = h.$$ (18)

Let us substitute (18) into (3)-(12), we receive the quanta of time, mass, length, volume (size) and charge:

$$T_q = \frac{G}{c^5} E_q = \frac{Gh}{c^5} = 1.82625 \cdot 10^{-86} \, s, \quad M_q = \frac{E_q}{c^2} = \frac{h}{c^2} = 7.37254 \cdot 10^{-51} \, kg,$$

$$R_q = \frac{G}{c^4} E_q = \frac{Gh}{c^4} = 5.47495 \cdot 10^{-78} \, m, \quad v_q = R_q^3 = \frac{G^3 h^3}{c^{12}} = 1.64112 \cdot 10^{-232} \, m^3,$$

$$Q_q = \left(\frac{hG^2 c}{k}\right)^{1/2} E_q = \left(\frac{h^3 G^2 c}{k}\right)^{1/2} = 2.079086 \cdot 10^{-61} \, C,$$ (19)

$$V_q = \frac{E_q}{Q_p} = \left(\frac{2\pi k}{c}\right)^{1/2} = 3.532876 \cdot 10^{-16} \, V, \quad I_q = \frac{Q_q}{T_p} = \left(\frac{2\pi h^2 c^6}{k}\right)^{1/2} = 3.85654 \cdot 10^{-18} \, A.$$

where v_q is quantum of volume, m³; V_q is quantum of voltage, V; I_q is quantum of the electric currency, A, Q_p, T_p are Planck units of charge and time respectively (see Table 2).

Heisenberg uncertainty principle

Heisenberg uncertainty principle are

$$\Delta I \cdot \Delta R \geq \hbar/2, \quad \Delta E \cdot \Delta T \geq \hbar/2, \quad \hbar = h/2\pi, \tag{20}$$

where ΔI, ΔR, ΔE, ΔT are uncertainty of momentum, length, energy and time respectively.

Substitute into (20) the quanta (19) we receive the following the uncertainties the main quanta (19)

$$\Delta E = \frac{h}{4\pi T_q} = \frac{c^5}{4\pi G} = 2.887272 \cdot 10^{51} \ J \quad for \quad T_q,$$

$$\Delta R = \frac{G}{c^4} \Delta E = \frac{c}{4\pi} = 2.38567258 \cdot 10^7 \ m \quad for \quad \Delta E \quad (T_q),$$

$$\Delta M = \frac{\Delta E}{c^2} = \frac{c^3}{4\pi G} = 3.212523 \cdot 10^{34} \ kg \quad for \quad \Delta E \quad (T_q), \tag{21}$$

$$\Delta Q = \left(\frac{G}{kc^4}\right)^{1/2} \Delta E = 8.6971819 \cdot 10^{24} \ kg \cdot m^2/s^2 \quad for \quad \Delta E \quad (T_q),$$

As you see, the uncertainties of quanta are big and we can not measure them. These values ΔE, ΔR, appears when are appeared in the first quantum of time T_q. The ΔM, ΔQ not appeared yet. They are equivalent the given ΔE.

The probability serve of inequality (20) is normal. If we take (20) in the more common form

$$\Delta I \cdot \Delta R \geq h, \quad \Delta E \cdot \Delta T \geq h, \tag{22}$$

the multiplier 4π in equations (21) become 1 and ΔR = c. That means the speed in the first quantum of time equals the light speed.

Note: For getting the values (3)-(21) we also used the dimension theory and some of them may be defined with accuracy the constant factor.

Main Results and Discussion

Main result of this research (part 1 -3) is equations with result that energy can be the universal source of Universe (see Eq.(5)). Energy can produce time, mass, charge and volume. The same role/factor also can act as time (see Eq. (3)). All main components of Universe (size, matter, energy, volume, time, charge) are closely connected and can transform from one to another.

That means at the foundation of the Universe is ONE factor, which creates our diverse world.

The reader can ask: How we can convert time to energy? I can ask a counter question: The equation $E = Mc^2$ (here M is mass) was open about hundred years ago. In that (past) time nobody could answer: How to convert the matter into this huge energy using this equation? Only tens of years later the scientists opened that certain nuclei of atoms can convert one to another, significantly change their mass and emit or absorb such quantity of energy. In 2006 the author offered the method which can convert any matter to energy with according to the equation $E = mc^2$ [5] – [6].

In Universe (Part 1)[1] author has developed a theory, which allows derivation of the unknown relations between main parameters (energy, time, volume, matter) in Universe. In given part 3 he added charge as main parameter in this theory. He finds also the quantum (minimal values) of energy, time, volume and matter and he applied these quantum for estimations of quantum volatility and the estimation of some values of our Universe and received both well-known and new unknown relations.

Only time and experiments can confirm, correct or deny the proposed formulae.

The authors other works closest to this topic are presented in references [1] – [7].

References:

1. Bolonkin A.A., Universe (part 1). Relations between Time, Matter, Volume, Distance, and Energy. JOURNAL OF ENERGY STORAGE AND CONVERSION, JESC : July-December 2012, Volume 3, #2, pp. 141-154. http://viXra.org/abs/1207.0075 , http://www.scribd.com/doc/100541327/ , http://archive.org/details/Universe.RelationsBetweenTimeMatterVolumeDistanceAndEnergy
2. Bolonkin A.A., Universe (Part 2): Rolling of Space (Volume, Distance, Time, and Matter) into a Point. http://www.scribd.com/doc/120693979
3. Bolonkin A.A., "Remarks about Universe" (part 1-2), International Journal of Advanced Engineering Applications, IJAEA. Vol.1, Iss.3, pp.62-67 (2012) . http://viXra.org/abs/1309.0196 , http://fragrancejournals.com/wp-content/uploads/2013/03/IJAEA-1-3-10.pdf
4. Bolonkin A.A., "New Technologies and Revolutionary Projects", Scribd, 2010, 324 pgs, http://www.scribd.com/doc/32744477 or http://www.archive.org/details/NewTechnologiesAndRevolutionaryProjects
5. Bolonkin A.A., Converting of Any Matter to Nuclear Energy by-AB-Generator American Journal of Engineering and Applied Science, Vol. 2, #4, 2009, pp.683-693. http://www.scribd.com/doc/24048466/
6. Bolonkin A.A., Universe, Human Immortality and Future Human Evaluation. Scribd. 2010r. 124 pages, 4.8 Mb. http://www.scribd.com/doc/52969933/ http://www.archive.org/details/UniverseHumanImmortalityAndFutureHumanEvaluation
7. AIP Physics Desk Reference, Springer. Wikipedia. Universe. http://Wikipedia.org .

30 December, 2013.

Chapter 3

Superstrong Nuclear Matter *

ABSTRACT

Aerospace, aviation particularly need, in any era, the strongest and most thermostable materials available, often at nearly any price. The Space Elevator, space ships (especially during atmospheric reentry), rocket combustion chambers, thermally challenged engine surfaces, hypersonic aircraft materials better than any now available, with undreamed of performance as the reward if obtained. As it is shown in this research, the offered new material allows greatly to improve the all characteristics of space ships, rockets, engines and aircraft and design new types space, propulsion, aviation systems.

At present the term 'nanotechnology' is well known – in its' ideal form, the flawless and completely controlled design of conventional molecular matter from molecules or atoms.

But even this yet unachieved goal is not the end of material science possibilities. The author herein offers the idea of design of new forms of nuclear matter from nucleons (neutrons, protons), electrons, and other nuclear particles. He shows this new 'AB-Matter' has extraordinary properties (for example, tensile strength, stiffness, hardness, critical temperature, superconductivity, supertransparency, zero friction, etc.), which are up to millions of times better than corresponding properties of conventional molecular matter. He shows concepts of design for space ships, rockets, aircraft, sea ships, transportation, thermonuclear reactors, constructions, and so on from nuclear matter. These vehicles will have unbelievable possibilities (e.g., invisibility, ghost-like penetration through any walls and armour, protection from nuclear bomb explosions and any radiation flux, etc.)

Nanotechnology, in near term prospect, operates with objects (molecules and atoms) having the size in nanometer (10^{-9} m). The author here outlines perhaps more distant operations with objects (nuclei) having size in the femtometer range, (10^{-15} m, millions of times less smaller than the nanometer scale). The name of this new technology is femtotechnology.

Key words: *femtotechnology, nuclear matter, artificial AB-Matter, superstrength matter, superthermal resistance, invisible matter, super-protection from nuclear explosion and radiation.*

* Presented as paper AIAA-2009-4620 to 45 Joint Propulsion Conference, 2-5 August, 2009, Denver CO, USA.

INTRODUCTION
Brief information concerning the atomic nucleus.

Atoms are the smallest (size is about some 10^{-8} m) neutral particles into which matter can be divided by chemical reactions. An atom consists of a small, heavy nucleus surrounded by a relatively large, light cloud of electrons. Each type of atom corresponds to a specific chemical element. To date, 117 elements have been discovered (atomic numbers 1-116 and 118), and the first 111 have received official names. The well-known periodic table provides an overview. Atoms consist of protons and neutrons within the nucleus. Within these particles, there are smaller particles still which are then made up of even smaller particles still.

Molecules are the smallest particles into which a non-elemental substance can be divided while maintaining the physical properties of the substance. Each type of molecule corresponds to a specific chemical compound. Molecules are a composite of two or more atoms.

Atoms contain small (size is about some 10^{-15} m) nuclei and electrons orbit around these nuclei. The nuclei of most atoms consist of protons and neutrons, which are therefore collectively referred to as nucleons. The number of protons in a nucleus is the atomic number and defines the type of element the atom forms. The number of neutrons determines the isotope of an element. For example, the carbon-12 isotope has 6 protons and 6 neutrons, while the carbon-14 isotope has 6 protons and 8 neutrons.

While bound neutrons in stable nuclei are stable, free neutrons are unstable; they undergo beta decay with a lifetime of just under 15 minutes. Free neutrons are produced in nuclear fission and fusion. Dedicated neutron sources like research reactors and spallation sources produce free neutrons for the use in irradiation and in neutron scattering experiments.

Outside the nucleus, free neutrons are unstable and have a mean lifetime of 885.7±0.8 s, decaying by emission of a negative electron and antineutrino to become a proton:

$$n^0 \rightarrow p^+ + e^- + \overline{v}_e.$$

This decay mode, known as beta decay, can also transform the character of neutrons within unstable nuclei.

Bound inside a nucleus, protons can also transform via inverse beta decay into neutrons. In this case, the transformation occurs by emission of a positron (antielectron) and a neutrino (instead of an antineutrino):

$$p^+ \rightarrow n^0 + e^+ + v_e.$$

The transformation of a proton to a neutron inside of a nucleus is also possible through electron capture:

$$p^+ + e^- \rightarrow n^0 + v_e.$$

Positron capture by neutrons in nuclei that contain an excess of neutrons is also possible, but is hindered because positrons are repelled by the nucleus, and quickly annihilate when they encounter negative electrons.

When bound inside of a nucleus, the instability of a single neutron to beta decay is balanced against the instability that would be acquired by the nucleus as a whole if an additional proton were to participate in repulsive interactions with the other protons that are already present in the nucleus. As such, although free neutrons are unstable, bound neutrons are not necessarily so. The same reasoning explains why protons, which are stable in empty space, may transform into neutrons when bound inside of a nucleus.

A thermal neutron is a free neutron that is Boltzmann distributed with kT = 0.024 eV (4.0×10^{-21} J) at room temperature. This gives characteristic (not average, or median) speed of 2.2 km/s.

Four forces active between particles: strong interaction, weak interacting, charge force (Coulomb force) and gravitation force. The strong interaction is the most strong force in short nuclei distance, the gravitation is very small into atom.

Beta decay and electron capture are types of radioactive decay and are both governed by the weak interaction.

Basic properties of the nuclear force.

The nuclear force is only felt among hadrons. In particle physics, a hadron is a bound state of quarks (particles into nucleous). Hadrons are held together by the strong force, similarly to how atoms are held together by the electromagnetic force. There are two subsets of hadrons: baryons and mesons; the most well known baryons are protons and neutrons.

At much smaller separations between nucleons the force is very powerfully repulsive, which keeps the nucleons at a certain average separation. Beyond about 1.7 femtometer (fm) separation, the force drops to negligibly small values.

At short distances, the nuclear force is stronger than the Coulomb force; it can overcome the Coulomb repulsion of protons inside the nucleus. However, the Coulomb force between protons has a much larger range and becomes the only significant force between protons when their separation exceeds about 2.5 fm.

The nuclear force is nearly independent of whether the nucleons are neutrons or protons. This property is called *charge independence*. It depends on whether the spins of the nucleons are parallel or antiparallel, and has a noncentral or *tensor* component. This part of the force does not conserve orbital angular momentum, which is a constant of motion under central forces.

The nuclear force (or **nucleon-nucleon interaction** or **residual strong force**) is the force between two or more nucleons. It is responsible for binding of protons and neutrons into atomic nuclei. To a large extent, this force can be understood in terms of the exchange of virtual light mesons, such as the pions. Sometimes the nuclear force is called the residual strong force, in contrast to the strong interactions which are now understood to arise from quantum chromodynamics (QCD). This phrasing arose during the 1970s when QCD was being established. Before that time, the *strong nuclear force* referred to the inter-nucleon potential. After the verification of the quark model, *strong interaction* has come to mean QCD.

A subatomic particle is an elementary or composite particle smaller than an atom. Particle physics and nuclear physics are concerned with the study of these particles, their interactions, and non-atomic matter.

Elementary particles are particles with no measurable internal structure; that is, they are not composed of other particles. They are the fundamental objects of quantum field theory. Many families and sub-families of elementary particles exist. Elementary particles are classified according to their spin. Fermions have half-integer spin while bosons have integer spin. All the particles of the Standard Model have been observed, with the exception of the Higgs boson.

Subatomic particles include the atomic constituents electrons, protons, and neutrons. Protons and neutrons are composite particles, consisting of quarks. A proton contains two up quarks and one down quark, while a neutron consists of one up quark and two down quarks; the quarks are held together in the nucleus by gluons. There are six different types of quark in all ('up', 'down', 'bottom', 'top', 'strange', and 'charm'), as well as other particles including photons and neutrinos which are produced copiously in the sun. Most of the particles that have been discovered are encountered in cosmic rays interacting with matter and are produced by scattering processes in particle accelerators. There are dozens of known subatomic particles.

Degenerate matter.

Degenerate matter is matter which has such very high density that the dominant contribution to its pressure rises from the Pauli exclusion principle. The pressure maintained by a body of degenerate matter is called the degeneracy pressure, and arises because the Pauli principle forbids the constituent particles to occupy identical quantum states. Any attempt to force them close enough together that they are not clearly separated by position must place them in different energy levels. Therefore, reducing the volume requires forcing many of the particles into higher-energy quantum states. This requires additional compression force, and is manifest as a resisting pressure.

Imagine that there is a plasma, and it is cooled and compressed repeatedly. Eventually, we will not be able to compress the plasma any further, because the Exclusion Principle states that two particles cannot be in the exact same place at the exact same time. When in this state, since there is no extra space for any particles, we can also say that a particle's location is extremely defined. Therefore, since (according to the Heisenberg Uncertainty Principle) $\Delta p \Delta x = h/2$ where Δp is the uncertainty in the particle's momentum and Δx is the uncertainty in position, then we must say that their momentum is extremely uncertain since the molecules are located in a very confined space. Therefore, even though the plasma is cold, the molecules must be moving very fast on average. This leads to the conclusion that if you want to compress an object into a very small space, you must use tremendous force to control its particles' momentum.

Unlike a classical ideal gas, whose pressure is proportional to its temperature ($PV = NkT$, where P is pressure, V is the volume, N is the number of particles (typically atoms or molecules), k is Boltzmann's constant, and T is temperature), the pressure exerted by degenerate matter depends only weakly on its temperature. In particular, the pressure remains nonzero even at absolute zero temperature. At relatively low densities, the pressure of a fully degenerate gas is given by
$P = Kn^{5/3}$, where K depends on the properties of the particles making up the gas. At very high densities, where most of the particles are forced into quantum states with relativistic energies, the pressure is given by $P = K'n^{4/3}$, where K' again depends on the properties of the particles making up the gas.

Degenerate matter still has normal thermal pressure, but at high densities the degeneracy pressure dominates. Thus, increasing the temperature of degenerate matter has a minor effect on total pressure until the temperature rises so high that thermal pressure again dominates total pressure.

Exotic examples of degenerate matter include neutronium, strange matter, metallic hydrogen and white dwarf matter. Degeneracy pressure contributes to the pressure of conventional solids, but these are not usually considered to be degenerate matter as a significant contribution to their pressure is provided by the interplay between the electrical repulsion of atomic nuclei and the screening of nuclei from each other by electrons allocated among the quantum states determined by the nuclear electrical potentials. In metals it is useful to treat the conduction electrons alone as a degenerate, free electron gas while the majority of the electrons are regarded as occupying bound quantum states. This contrasts with the case of the degenerate matter that forms the body of a white dwarf where all the electrons would be treated as occupying free particle momentum states.

Pauli principle

The Pauli exclusion principle is a quantum mechanical principle formulated by Wolfgang Pauli in 1925. It states that no two identical fermions may occupy the same quantum state *simultaneously*. A more rigorous statement of this principle is that, for two identical fermions, the total wave function is anti-symmetric. For electrons in a single atom, it states that no two electrons can have the same four quantum numbers, that is, if n, l, and m_l are the same, m_s must be different such that the electrons have opposite spins.

In relativistic quantum field theory, the Pauli principle follows from applying a rotation operator in imaginary time to particles of half-integer spin. It does not follow from any spin relation in nonrelativistic quantum mechanics.

The Pauli exclusion principle is one of the most important principles in physics, mainly because the three types of particles from which ordinary matter is made—electrons, protons, and neutrons—are all subject to it; consequently, all material particles exhibit space-occupying behavior. The Pauli exclusion principle underpins many of the characteristic properties of matter from the large-scale stability of matter to the existence of the periodic table of the elements. Particles with antisymmetric wave functions are called fermions—and obey the Pauli exclusion principle. Apart from the familiar electron, proton and neutron, these include neutrinos and quarks (from which protons and neutrons are made), as well as some atoms like helium-3. All fermions possess "half-integer spin", meaning that they possess an intrinsic angular momentum whose value is $\Delta p \Delta x \geq h/2$ (Planck's constant divided by 2π) times a half-integer (1/2, 3/2, 5/2, etc.). In the theory of quantum mechanics, fermions are described by "antisymmetric states", which are explained in greater detail in the theory on identical particles. Particles with integer spin have a symmetric wave function and are called bosons; in contrast to fermions, they may share the same quantum states. Examples of bosons include the photon, the Cooper pairs responsible for superconductivity, and the W and Z bosons.

A more rigorous proof was provided by Freeman Dyson and Andrew Lenard in 1967, who considered the balance of attractive (electron-nuclear) and repulsive (electron-electron and nuclear-nuclear) forces and showed that ordinary matter would collapse and occupy a much smaller volume without the Pauli principle.

Neutrons are the most "rigid" objects known - their Young modulus (or more accurately, bulk modulus) is 20 orders of magnitude larger than that of diamond.

For white dwarfs the degenerate particles are the electrons while for neutron stars the degenerate particles are neutrons. In degenerate gas, when the mass is increased, the pressure is increased, and the particles become spaced closer together, so the object becomes smaller. Degenerate gas can be compressed to very high densities, typical values being in the range of 10^7 grams per cubic centimeter.

Preons are subatomic particles proposed to be the constituents of quarks, which become composite particles in preon-based models.

Neutron stars.

A neutron star is a large gravitationally-bound lump of electrically neutral nuclear matter, whose pressure rises from zero (at the surface) to an unknown value in the center.

A neutron star is a type of remnant that can result from the gravitational collapse of a massive star during a Type II, Type Ib or Type Ic supernova event. Such stars are composed almost entirely of neutrons, which are subatomic particles with zero electrical charge and roughly the same mass as protons.

A typical neutron star has a mass between 1.35 and about 2.1 solar masses, with a corresponding radius of about 12 km if the Akmal-Pandharipande-Ravenhall (APR) Equation of state (EOS) is used. In contrast, the Sun's radius is about 60,000 times that. Neutron stars have overall densities predicted by the APR EOS of 3.7×10^{17}

(2.6×10^{14} times Solar density) to 5.9×10^{17} kg/m³ (4.1×10^{14} times Solar density). which compares with the approximate density of an atomic nucleus of 3×10^{17} kg/m³. The neutron star's density varies from below 1×10^{9} kg/m³ in the crust increasing with depth to above 6 or 8×10^{17} kg/m³ deeper inside.

In general, compact stars of less than 1.44 solar masses, the Chandrasekhar limit, are white dwarfs; above 2 to 3 solar masses (the Tolman-Oppenheimer-Volkoff limit), a quark star might be created, however this is uncertain. Gravitational collapse will always occur on any star over 5 solar masses, inevitably producing a black hole.

The gravitational field at the star's surface is about 2×10^{11} times stronger than on Earth. The escape velocity is about 100,000 km/s, which is about one third the speed of light. Such a strong gravitational field acts as a gravitational lens and bends the radiation emitted by the star such that parts of the normally invisible spectrum near the surface become visible.

The gravitational binding energy of a two solar mass neutron star is equivalent to the total conversion of one solar mass to energy (From the law of mass-energy equivalence, $E=mc^2$). That energy was released during the supernova explosion.

A neutron star is so dense that one teaspoon (5 millilitre) of its material would have a mass over 5×10^{12} kg. The resulting force of gravity is so strong that if an object were to fall from just one meter high it would hit the surface of the neutron star at 2 thousand kilometers per second, or 4.3 million miles per hour.

The Equation of state (EOS) for a Neutron star is still not known as of 2008.

On the basis of current models, the matter at the surface of a neutron star is composed of ordinary atomic nuclei as well as electrons.

INNOVATIONS AND COMPUTATIONS

1. **Short information about atom and nuclei**. Conventional matter consists of atoms and molecules. Molecules are collection of atoms. The atom contains a nucleus with proton(s) and usually neutrons (Except for Hydrogen-1) and electrons revolve around this nucleus. Every particle may be characterized by parameters as mass, charge, spin, electric dipole, magnetic moment, etc. There are four forces active between particles: strong interaction, weak interaction, electromagnetic charge (Coulomb) force and gravitational force. The nuclear force dominates at distances up to 2 fm (femto, 1 fm = 10^{-15} m). They are hundreds of times more powerful than the charge (Coulomb force and million-millions of times more than gravitational force. Charge (Coulomb) force is effective at distances over 2 fm. Gravitational force is significant near and into big masses (astronomical objects such as planets, stars, white dwarfs, neutron stars and black holes). Strong force is so overwhelmingly powerful that it forces together the positively charged protons, which would repel one from the other and fly apart without it. The strong force is key to the relationship between protons, neutrons and electrons. They can keep electrons into or near nuclei. Scientists conventionally take into attention only of the strong force when they consider the nuclear and near nuclear size range, for the other forces on that scale are negligible by comparison for most purposes.

Strong nuclear forces are anisotropic (non spherical, force distribution not the same in all directions equally), which means that they depend on the relative orientation of the nucleus.

Typical nuclear energy (force) is presented in fig.1. When it is positive the nuclear force repels the other atomic particles (protons, neutrons, electrons). When nuclear energy is negative, it attracts them up to a distance of about 2 fm. The value r_0 usually is taken as radius of nucleus. The computation of strong nuclear force - interaction energy of one nucleus via specific density of one nucleus in given point – is present in Fig.2. The solid line is as computed by Berkner's method [7] with 2 correlations, dotted line is computer generated with 3 correlations, square is experimental. Average interaction energy between to nucleus is about 8 MeV, distance where the attractive strong nuclear force activates is at about 1 – 1.2 fm.

2. **AB-Matter**. In conventional matter made of atoms and molecules the nucleons (protons, neutrons) are located in the nucleus, but the electrons rotate in orbits around nucleus in distance in millions times more than diameter of nucleus. Therefore, in essence, what we think of as solid matter contains a -- relatively! --'gigantic'

vacuum (free space) where the matter (nuclei) occupies but a very small part of the available space. Despite this unearthly emptiness, when you compress this (normal, non-degenerate) matter the electrons located in their orbits repel atom from atom and resist any great increase of the matter's density. Thus it feels solid to the touch.

The form of matter containing and subsuming all the atom's particles into the nucleus is named *degenerate matter*. Degenerate matter found in white dwarfs, neutron stars and black holes. Conventionally this matter in such large astronomical objects has a high temperature (as independent particles!) and a high gravity adding a forcing, confining pressure in a very massive celestial objects. In nature, degenerate matter exists stably (as a big lump) to our knowledge only in large astronomical masses (include their surface where gravitation pressure is zero) and into big nuclei of conventional matter.

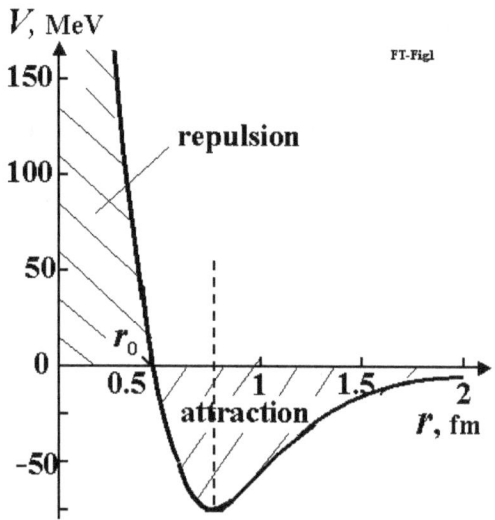

Fig.1. Typical nuclear force of nucleus. When nucleon is at distance of less than 1.8 fm, it is attracted to nucleus. When nucleon is very close, it is repulsed from nucleus. (Reference from http://www.physicum.narod.ru , Vol. 5 p. 670).

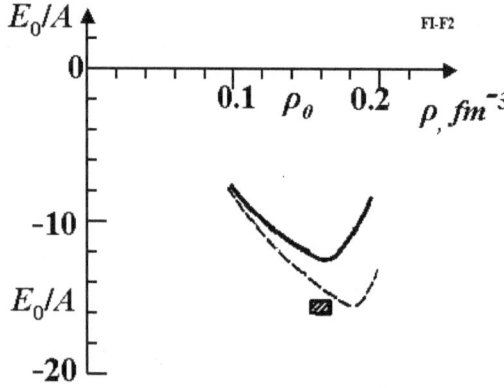

Fig.2. Connection (interaction) energy of one nucleon via specific density of one nucleon in given point. Firm line is computed by Berkner's method with 2 correlations, dotted line is computer with 3 correlations, square is experiment. (Reference from http://www.physicum.narod.ru , Vol. 5 p. 655).

Our purpose is to design artificial small masses of synthetic degenerate matter in form of an extremely thin strong thread (fiber, filament, string), round bar (rod), tube, net (dense or non dense weave and mesh size) which can exist at Earth-normal temperatures and pressures. Note that such stabilized degenerate matter in small amounts does not exist in Nature as far as we know. Therefore author has named this matter **AB-Matter**.

Just as people now design by the thousands variants of artificial materials (for example, plastics) from usual matter, we soon (historically speaking) shall create many artificial, designer materials by nanotechnology (for example, nanotubes: SWNTs (amchair, zigzag, ahiral), MWNTs (fullorite, torus, nanobut), nanoribbon (plate), buckyballs (ball), fullerene). Sooner or later we may anticipate development of femtotechnology and create such AB-Matter. Some possible forms of AB-Matter are shown in fig.3. Offered technologies are below. The threads from AB-Matter are stronger by millions of times than normal materials. They can be inserted as reinforcements, into conventional materials, which serve as a matrix, and are thus strengthened by thousands of times (see computation section).

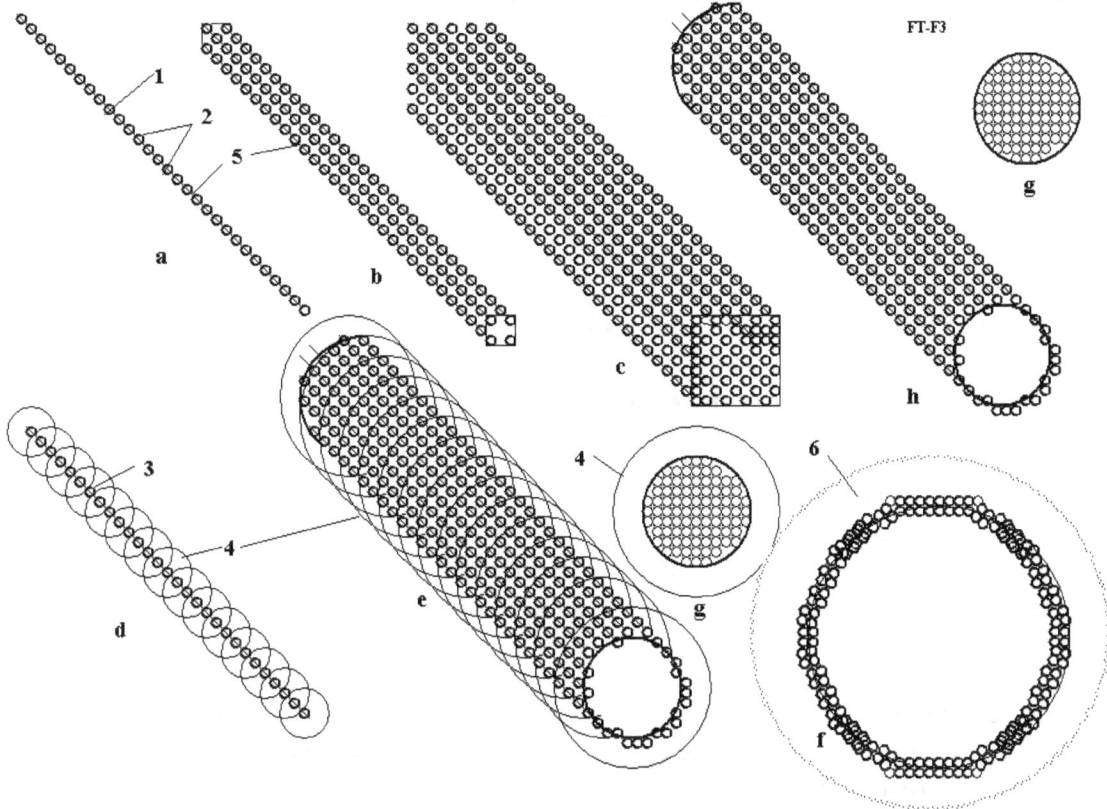

Fig.3. Design of AB-Matter from nucleons (neutrons, protons, etc.) and electrons (**a**) linear one string (monofilament) (fiber, whisker, filament, thread); (**b**) ingot from four nuclear monofilaments; (**c**) multi-ingot from nuclear monofilament; (**d**) string made from protons and neutrons with electrons rotated around monofilament; (**e**) single wall femto tube (SWFT) fiber with rotated electrons; (**f**) cross-section of multi wall femto tube (MWFT) string; (**g**) cross-section of rod; (**h**) - single wall femto tube (SWFT) string with electrons inserted into AB-Matter. *Notations*: 1 – nuclear string; 2 - nucleons (neutrons, protons, etc.). 3 – protons; 4 – orbit of electrons; 5 – electrons; 6 – cloud of electrons around tube.

3. Some offered technologies for producing AB-Matter. One method of producing AB-Matter may use the technology reminiscent of computer chips (fig.4). One side of closed box 1 is evaporation mask 2. In the other size are located the sources of neutrons, charged nuclear particles (protons, charged nuclei and their connections) and electrons. Sources (guns) of charged particles have accelerators of particles and control their energy and direction. They concentrate (focus) particles, send particles (in beam form) to needed points with needed energy for overcoming the Coulomb barrier. The needed neutrons are received also from nuclear reactions and reflected by the containing walls.

Various other means are under consideration for generation of AB-Matter, what is certain however is that once the first small amounts have been achieved, larger and larger amounts will be produced with ever increasing ease. Consider for example, that once we have achieved the ability to make a solid AB-Matter film (a sliced plane through a solid block of AB-matter), and then developed the ability to place holes with precision through it one nucleon wide, a modified extrusion technique may produce AB-Matter strings (thin fiber), by passage of conventional matter in gas, liquid or solid state through the AB-Matter matrix (mask). This would be a 'femto-die' as Joseph Friedlander of Shave Shomron, Israel, has labeled it. Re-assembling these strings with perfect precision and alignment would produce more AB-matter film; leaving deliberate gaps would reproduce the 'holes' in the initial 'femto-die'.

The developing of femtotechnology is easier, in one sense, than the developing of fully controllable nanotechnology because we have only three main particles (protons, neutrons, their ready combination of nuclei $_2D$, $_3T$, $_4He$, etc., and electrons) as construction material and developed methods of their energy control, focusing and direction.

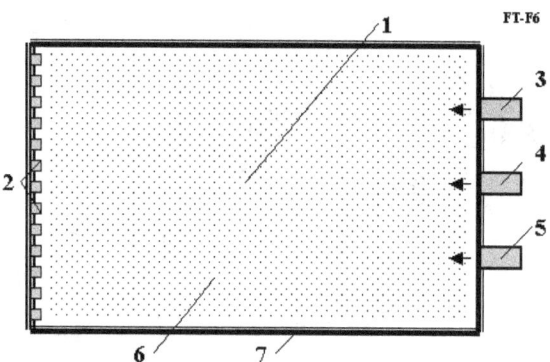

Fig.4. Conceptual diagram for installation producing AB-Matter. Notations: 1 – installation; 2 –AB-Matter (an extremely thin thread, round bar, rod, tube, net) and form mask; 3 – neutron source; 4 – source of charged particles (protons, charged nuclei), accelerator of charged particle, throttle control, beam control; 5 - source of electrons, accelerator of electrons, throttle control, beam control; 6 – cloud of particles; 7 – walls reflect the neutrons and utilize the nuclear energy.

4. Using the AB-Matter. The simplest use of AB-Matter is strengthening and reinforcing conventional material by AB-Matter fiber. As it is shown in the 'Computation' section, AB-Matter fiber is stronger (has a gigantic ultimate tensile stress) than conventional material by a factor of millions of times, can endure millions degrees of temperature, don't accept any attacking chemical reactions. We can insert (for example, by casting around the reinforcement) AB-Matter fiber (or net) into steel, aluminum, plastic and the resultant matrix of conventional material increases in strength by thousands of times—if precautions are taken that the reinforcement stays put! Because of the extreme strength disparity design tricks must be used to assure that the fibers stay 'rooted'. The matrix form of conventional artificial fiber reinforcement is used widely in current technology. This increases the tensile stress resistance of the reinforced matrix matter by typically 2 – 4 times. Engineers dream about a nanotube reinforcement of conventional matrix materials which might increase the tensile stress by 10 – 20 times, but nanotubes are very expensive and researchers cannot decrease its cost to acceptable values yet despite years of effort.

Another way is using a construct of AB-Matter as a continuous film or net (fig. 5b,d).

These forms of AB-Matter have such miraculous properties as invisibility, superconductivity, zero friction, etc. The ultimate in camouflage, installations of a veritable Invisible World can be built from certain forms of AB-Matter with the possibility of being also interpenetable, literally allowing ghost-like passage through an apparently solid wall. Or the AB-Matter net (of different construction) can be designed as an impenetrable wall that even hugely destructive weapons cannot penetrate.

The AB-Matter film and net may be used for energy storage which can store up huge energy intensities and used also as rocket engines with gigantic impulse or weapon or absolute armor (see computation and application sections). Note that in the case of absolute armor, safeguards must be in place against buffering sudden accelerations; g-force shocks can kill even though nothing penetrates the armor!

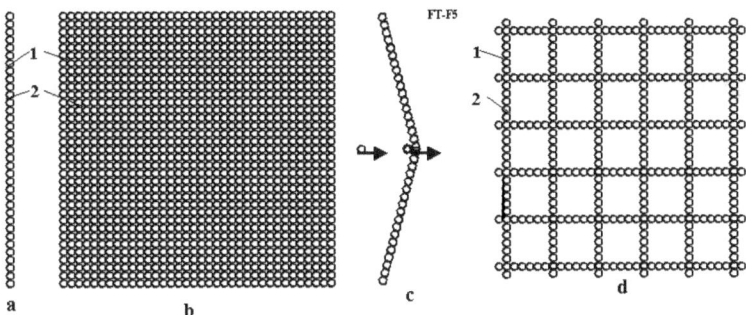

Fig.5. Thin film from nuclear matter. (**a**) cross-section of a matter film from single strings (side view); (**b**) continuous film from nuclear matter; (**c**) AB film under blow from conventional molecular matter; (**d**) – net from single strings. Notations: 1 – nucleons; 2 – electrons inserted into AB-Matter; 3 – conventional atom.

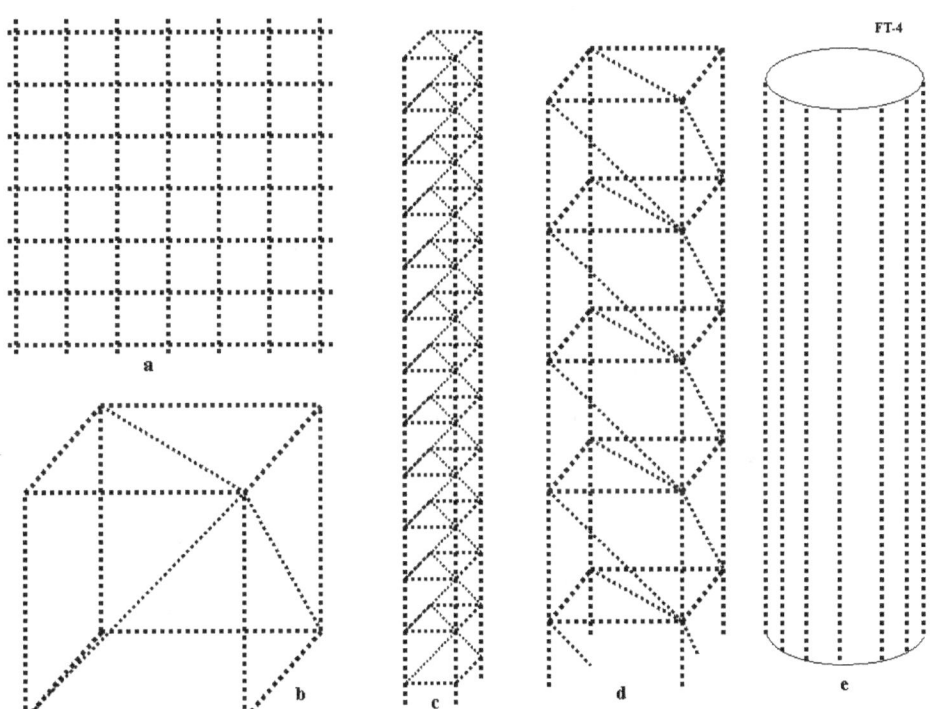

Fig.6. Structures from nuclear strings. (**a**) nuclear net (netting, gauze); (**b**) primary cube from matter string; (**c**) primary column from nuclear string; (**d**) large column where elements made from primary columns; (**e**) tubes from matter string or matter columns.

The AB-Matter net (which can be designed to be gas-impermeable) may be used for inflatable construction of such strength and lightness as to be able to suspend the weight of a city over a vast span the width of a sea. AB-Matter may also be used for cubic or tower solid construction as it is shown in fig.6.

ESTIMATION AND COMPUTATION OF PROPERTIES OF AB-MATTER

1. Strength of AB-Matter.

Strength (tensile stress) of single string (AB-Matter monofilament). The average connection energy of two nucleons is

$$1 \text{ eV} = 1.6 \times 10^{-19} \text{ J}, \quad E = 8 \text{ MeV} = 12.8 \times 10^{-13} \text{ J}. \tag{1}$$

The average effective distance of the strong force is about $l = 2$ fm $= 2 \times 10^{-15}$ m (1 fm $= 10^{-15}$ m). The average connection force F the single thread is about

$$F_1 = E/l = 6.4 \times 10^2 \text{ N}. \tag{2}$$

This is worth your attention: a thread having diameter *100 thousand times less than an atom's diameter* can suspend a weight nearly of human mass. The man may be suspended this invisible and permeable thread(s) and people will not understand how one fly.

Specific ultimate tensile stress of single string for cross-section area $s = 2 \times 2 = 4$ fm^2 $= 4 \times 10^{-30}$ m^2 is

$$\sigma = F/s = 1.6 \times 10^{32} \text{ N/m}^2. \tag{3}$$

Compressive stress for $E = 30$ MeV and $l = 0.4$ fm (fig.1) is

$$\sigma = E/sl = 3 \times 10^{33} \text{ N/m}^2. \tag{4}$$

The Young's modulus of tensile stress for elongation of break $\varepsilon = 1$ is

$$l = \sigma/\varepsilon = 1.6 \times 10^{32} \text{ N/m}^2. \tag{5}$$

The Young's modulus of compressive stress for $\varepsilon = 0.4$ is

$$l = \sigma/\varepsilon = 7.5 \times 10^{33} \text{ N/m}^2. \tag{6}$$

Comparison: Stainless steel has a value of $\sigma = (0.65 - 1) \times 10^9$ N/m^2, $l = 2 \times 10^{11}$ N/m^2. Nanotubes has $\sigma = (1.4 \div 5) \times 10^{10}$ N/m^2, $l = 8 \times 10^{11}$ N/m^2. That means AB-Matter is stronger by a factor of 10^{23} times than steel (by 100 thousands billion by billions times!) and by 10^{22} times than nanotubes (by 10 thousand billion by billions times!). Young's modulus, and the elastic modulus also are billions of times more than steel and elongation is tens times better than the elongation of steel.

Strength (average tensile force) of one m thin (one layer, 1 fm) film (1 m compact net) from single strings with step size of grid $l = 2$ fm $= 2 \times 10^{-15}$ m is

$$F = F_1/l = 3.2 \times 10^{17} \text{ N/m} = 3.2 \times 10^{13} \text{ tons/m}. \tag{7}$$

Strength (average tensile force) of net from single string with step (mesh) size $l = 10^{-10}$ m (less than a molecule size of conventional matter) which does not pass the any usual gas, liquids or solid (an impermeable net, essentially a film to ordinary matter)

$$F = F_1/l = 6.4 \times 10^{12} \text{ N/m} = 6.4 \times 10^8 \text{ tons/m}. \tag{8}$$

That means one meter of very thin (1 fm) net can suspend *100 millions tons of load.*

The tensile stress of a permeable net (it will be considered later) having $l = 10^{-7}$ m is

$$F = F_1/l = 6.4 \times 10^9 \text{ N/m} = 6.4 \times 10^5 \text{ tons/m}. \tag{9}$$

2. Specific density and specific strength of AB-Matter.

The mass of 1 m of single string (AB-Matter. Monofilament) is

$$M_1 = m/l = 1.67 \times 10^{-27}/(2 \times 10^{-15}) = 8.35 \times 10^{-13} \text{ kg}. \tag{10}$$

where $m = 1.67 \times 10^{-27}$ kg is mass of one nucleon; $l = 2 \times 10^{-15}$ is distance between nucleons, m., the volume of 1 m one string is $v = 10^{-30}$ m^3. That means the specific density of AB-Matter string and compact net is

$$d = \gamma = M_1/v = 8.35 \times 10^{17} \text{ kg/m}^3. \tag{11}$$

That is very high (nuclear) specific density. But the total mass is nothing to be afraid of since, the dimensions of AB-Matter string, film and net are very small and mass of them are:

a) mass of string $M_1 = 8.35 \times 10^{-13}$ kg (see (10)), (12)
b) mass of 1 m^2 solid film $M_f = M_1/l = 4.17 \times 10^2$ kg, $l = 2 \times 10^{-15}$. (13)
c) mass of 1 m^2 impenetrable net $M_i = M_1/l = 8.35 \times 10^{-3}$ kg, $l = 10^{-10}$ m, (14)
d) mass of 1 m^2 permeable net $M_p = M_1/l = 8.35 \times 10^{-6}$ kg, $l = 10^{-7}$ m. (15)

As you see the fiber, nets from **AB-Matter** have very high strength and very small mass. To provide an absolute heat shield for the Space Shuttle Orbiter that could withstand reentries dozens of times worse than today would take only ~100 kilograms of mass for 1105 square meters of surface and the offsetting supports.

The specific strength coefficient of AB-Matter-- very important in aerospace-- [3]-]5] is

$k = \sigma/d = 1.6 \times 10^{32} /8.35 \times 10^{17} =1.9 \times 10^{14}$ (m/s)$^2 < c^2 = (3 \times 10^8)^2 = 9 \times 10^{16}$ (m/s)^2. (16)

This coefficient from conventional high strong fiber has value about $k = (1 - 6) \times 10^9$ [3]-[6].
AB-Matter is 10 million times stronger.

The specific mass and volume density of energy with AB-Matter are

$$E_v = E/v = 1.6 \times 10^{32} \text{ J/m}^3, \quad E_m = E/m_p = 7.66 \times 10^{14} \text{ J/kg}. \quad (17)$$

Here $E = 12.8 \times 10^{-13}$ J is (1), $m_p = 1.67 \times 10^{-27}$ kg is nucleon mass, kg, $v = 8 \times 10^{-45}$ m^3 is volume of one nucleon. The average specific pressure may reach

$$p = F_1/s = 12.8 \times 10^{-13} / 4 \times 10^{-30} = 3.2 \times 10^{-27} \text{ N/m}^2.$$

3. Failure temperature of AB-Matter and suitability for thermonuclear reactors.

The strong nuclear force is very powerful. That means the outer temperature which must to be reached to destroy the AB fiber, film or net is T_e = 6 MeV. If we transfer this temperature in Kelvin degrees we get

$$T_k = 1.16 \times 10^4 \, T_e = 7 \times 10^{10} \text{K}. \quad (18)$$

That temperature is 10 thousands millions degrees. It is about 50 - 100 times more than temperature in a fusion nuclear reactor. The size and design of the fusion reactor may be small and simple (for example, without big superconductive magnets, cryogenics, etc). We can add the AB matter has zero heat/thermal conductivity (see later) and it cannot cool the nuclear plasma. This temperature is enough for nuclear reaction of the cheap nuclear fuel, for example, D + D. The AB matter may be used in a high efficiency rocket and jet engines, in a hypersonic aircraft and so on.

No even in theory can conventional materials have this fantastic thermal resistance!

4. Energy generated by production of AB-Matter.

Getting of AB-matter produces a large amount of nuclear energy. That energy is more than the best thermonuclear fusion reaction produces. Joining of each nucleon produces 8 MeV energy, when joining the deuterium D and tritium T (2+3=5 nucleolus) produced only 17.5 MeV (3.6 MeV for every nucleon). If we use the ready blocks of nucleons as the D=^{2}H, T=^{3}H, ^{4}He, etc., the produced energy decreases. Using the ready nucleus blocks may be necessary because these reactions create the neutrons (n). For example:

$$^2H + {}^2H \rightarrow {}^3He + n + 3.27 \text{ MeV}, \quad {}^3H + {}^2H \rightarrow {}^4He + n + 17.59 \text{ MeV}, \quad (19)$$

which may be useful for producing the needed AB-Matter.

Using the ready blocks of nucleons decreases the energy getting in AB-Matter production but that decreases also the cost of needed material and enormously simplifies the technology.

A small part (0.7 MeV) of this needed energy will be spent to overcome the Coulomb barrier when the proton joins to proton. Connection of neutrons to neutron or proton does not request this energy (as there is no repulsion of charges). It should be no problem for current technology to accelerate the protons for energy 0.7 MeV.

For example, compute the energy in production of m = 1 gram = 0.001 kg of AB-Matter.

$$E_{1g} = E_1 m/m_p = 7.66 \times 10^{11} \text{ J/g}. \quad (20)$$

Here E_1= 8 MeV= 12.8×10^{-13} J – energy produced for joining 1 nucleon, $m_p = 1.67 \times 10^{-27}$ kg is mass of nucleon.

One kg of gasoline (benzene) produces 44 MJ/kg energy. That means that 1 g of AB-Matter requires the equivalent energy of 17.4 tons of benzene.

5. Super-dielectric strength of AB-Matter film. Dielectric strength equals

$$E_d = E/l = 8 \text{ MV}/10^{-15} \text{ m} = 8 \times 10^{15} \text{ MV/m}. \quad (21)$$

The best conventional material has dielectric strength of only 680 MV/m [4].

6. AB-Matter with orbiting electrons or immersed in electron cloud.

We considered early the AB-Matter which contains the electrons within its' own string, film or net. The strong nuclear force keeps the electron (as any conventional matter particle would) in its sphere of influence. But another method of interaction and compensation of electric charges is possible– rotation of electrons around AB-Matter string (or other linear member) or immersing the AB-Matter string (or other linear member, or AB-Matter net --) in a sea of electrons or negative charged atoms (ions). The first case is shown in fig. 3d,e,g , the second case is shown in fig. 3f.

The first case looks like an atom of conventional matter having the orbiting electron around the nucleus. However our case has a principal difference from conventional matter. In normal matter the electron orbits around the nucleus as a POINT. In our case it orbits around the charged nuclear material (AB-Matter) LINE (some form of linear member from AB-Matter). That gives a very important difference in electrostatic force acting on the electron. In conventional cases (normal molecular matter) the electrostatic force decreases as $1/r^2$, in our AB-Matter case the electrostatic force decreases as $1/r$. The interesting result (see below) is that the electron orbit in AB-Matter does follow the usual speed relationship to radius. The proof is below:

$$\frac{mV^2}{r} = eE, \quad E = k\frac{2\tau}{r}, \quad mV^2 = 2k\tau e, \quad V = \sqrt{\frac{2ke\tau}{m}} = \sqrt{N_p}e\sqrt{\frac{2k}{m}} = 22.4\sqrt{N_p}, \quad (22)$$

where $m = m_e = 9.11 \times 10^{-31}$ kg; V – electron speed, m/s; r is radius of electron orbit, m; τ is charge density in 1 m of single string, C/m; E is electrostatic intensity, A/m or N/C; $k = 9 \times 10^9$ Nm^2/C^2 is electrostatic constant, $e = 1.6 \times 10^{-19}$ C is charge of electron, C; N_p is number of proton in 1 m of single string, 1/m. As you see from last equation (22) the electron speed is not relative to radius. The real speed will be significantly less than given equation (22) because the other electrons block the charge of the rest of the string.

The total charge of the system is zero. Therefore we can put $N_p = 1$ (every electron in orbit is kept by only one proton in string). From last equation (22) we find $V = 22.4$ m/s. That means the electron speed carries only a very small energy.

In the second case the AB-Matter (string girder) can swim in a cloud (sea) of electrons. That case occurs in metals of conventional matter. But a lattice of metallic ions fills the volume of conventional metal giving drag to electron flow (causing electrical resistance).

The stringers and plate nets of AB-Matter can locate along the direction of electric flow. They constitute only a relatively tiny volume and will produce very small electric resistance. That means the AB-Matter may be quasi-super-conductivity or super-conductivity.

The electrons rotate around an AB-Matter string repel one from other. The tensile force from them is

$$F = k\frac{e^2}{d^2}\left(1 + \frac{1}{2^2} + \frac{1}{3^2} + \ldots + \frac{1}{n^2} + \ldots\right) = \frac{\pi^2 k}{6}\frac{e^2}{d^2} = 1.476 \cdot 10^{10}\frac{e^2}{d^2}. \quad (23)$$

For distance $d = 2 \times 10^{-15}$ m the force equals $F = 10.5$ N. This force keeps the string and net in unfolded stable form.

SOME PROPERTIES OF AB-MATTER

We spoke about the *fantastic tensile and compressive strength, rigidity, hardness, specific strength, thermal (temperature) durability, thermal shock, and big elongation of* AB-Matter.

Short note about other miraculous AB-Matter properties:

1. *Zero heat/thermal capacity*. That follows because the mass of nucleons (AB-Matter string, film, net) is large in comparison with mass single atom or molecule and nucleons in AB-Matter have a very strong connection one to other. Conventional atoms and molecules cannot pass their paltry energy to AB-Matter! That would be equivalent to moving a huge dry-dock door of steel by impacting it with very light table tennis balls.

2. *Zero heat/thermal conductivity*. (See above).

3. *Absolute chemical stability. No corrosion, material fatigue. Infinity of lifetime*. All chemical reactions are acted through ORBITAL electron of atoms. The AB-Matter does not have orbital electrons (special cases will be considered later on). Nucleons cannot combine with usual atoms having electrons. In particular, the AB-Matter has *absolute corrosion resistance. No fatigue of material* because in conventional material fatigue is result of splits between material crystals. No crystals in AB-Matter. That means AB-Matter has lifetime equal to the lifetime of neutrons themselves. Finally a container for the universal solvent!

4. *Super-transparency, invisibility of special AB-Matter-nets.* An AB-Matter net having a step distance (mesh size) between strings or monofilaments of more than 100 fm = 10^{-13} m will pass visible light having the wave length (400 - 800)×10^{-9} m. You can make cars, aircraft, and space ships from such a permeable (for visible light) AB-Matter net and you will see a man (who is made from conventional matter) apparently sitting on nothing, traveling with high speed in atmosphere or space without visible means of support or any visible vehicle!

5. *Impenetrability for gas, liquids, and solid bodies.* When the AB-Matter net has a step size between strings of less than atomic size of 10^{-10} m, it became impenetrabile for conventional matter. Simultaneously it may be invisible for people and have gigantic strength. The AB-Matter net may –as armor--protect from gun, cannon shells and missiles.

6. *Super-impenetrability for radiation.* If the cell size of the AB-Matter net will be less than a wave length of a given radiation, the AB-Matter net does not pass this radiation. Because this cell size may be very small, AB net is perfect protection from any radiation up to soft gamma radiation (include radiation from nuclear bomb).

7. *Full reflectivity (super-reflectivity).* If the cell size of an AB-Matter net will be less than a wavelength of a given radiation, the AB-Matter net will then fully reflect this radiation. With perfect reflection and perfect impenetrability remarkable optical systems are possible. A Fresnel like lens might also be constructible of AB-Matter.

8. *Permeable property (ghost-like intangibility power; super-passing capacity).* The AB-Matter net from single strings having mesh size between strings of more than 100 nm = 10^{-11} m will pass the atoms and molecules through itself because the diameter of the single string (2×10^{-15} m) is 100 thousand times less then diameter of atom (3×10^{-10} m). That means that specifically engineered constructions from AB-Matter can be built on the Earth, but people will not see and feel them. The power to phase through walls, vaults, and barriers has occasionally been portrayed in science fiction but here is a real life possibility of it happening.

9. *Zero friction.* If the AB-Matter net has a mesh size distance between strings equals or less to the atom (3×10^{-10} m), it has an ideal flat surface. That means the mechanical friction may be zero. It is very important for aircraft, sea ships and vehicles because about 90% of its energy they spend in friction. Such a perfect surface would be of vast value in optics, nanotech molecular assembly and prototyping, physics labs, etc.

10. *Super or quasi-super electric conductivity at any temperature.* As it is shown in previous section the AB-Matter string can have outer electrons in an arrangement similar to the electronic cloud into metal. But AB-Matter strings (threads) can be located along the direction of the electric intensity and they will not resist the electron flow. That means the electric resistance will be zero or very small.

11. *High dielectric strength* (see (21)).

AB-Matter may be used for devices to produce high magnetic intensity.

APPLICATIONS AND NEW SYSTEMS IN AEROSPACE AND AVIATION

The applications of the AB-Matter are encyclopedic in scope. This matter will create revolutions in many fields of human activity. We show only non-usual applications in aerospace, aviation that come to mind, and by no means all of these.

1. Storage of gigantic energy.

As it is shown in [3]-[7], the energy saved by flywheel equals the special mass density of material (17). As you see that is a gigantic value of stored energy because of the extreme values afforded by the strong nuclear force. Car having a pair of 1 gram counterspun fly-wheels (2 grams total) (20) charged at the factory can run all its life without benzene. Aircraft or sea ships having 100 gram (two 50 gram counterspun fly-wheels) can fly or swim all its life without additional fuel. The offered flywheel storage can has zero friction and indefinite energy storage time.

2. New propulsion system of space ship.

The most important characteristic of rocket engine is specific impulse (speed of gas or other material flow out from propulsion system). Let us compute the speed of a part of fly-wheel ejected from the offered rocket system

$$\frac{mV^2}{2} = E, \quad V = \sqrt{\frac{2E}{m}} = 3.9 \cdot 10^7 \ m/s. \tag{24}$$

Here V is speed of nucleon, m/s; $E = 12.8 \times 10^{-13}$ J (1) is energy of one nucleon, J; $m = 1,67 \times 10^{-27}$ kg is mass of one nucleon, kg. The value (24) is about 13% of light speed.

The chemical rocket engine has specific impulse about 3700 m/s. That value is 10 thousand times less. The electric rocket system has a high specific impulse but requires a powerful compact and light source of energy. In the offered rocket engine the energy is saved in the flywheel. The current projects of a nuclear rocket are very complex, heavy, and dangerous for men (gamma and neutron radiation) and have specific impulse of thousand of times less (24). The offered AB-Matter rocket engine may be very small and produced any rocket thrust in any moment in any direction.

The offered flywheel rocket engine used the AB-matter is presented in fig.7a. That is flywheel made from AB-matter. It has a nozzle 3 having control of exit mass. The control allows to exit of work mass in given moment and in given position of flywheel. The flywheel rotates high speed and the exhaust mass leave the rocket engine with same speed when the nozzle is open. In result the engine has thrust 6. As exhaust mass may be used any mass: liquid (for example, water), sand, small stones and other suitable planet or space material (mass). The energy needed for engine and space ship is saved in the revolving flywheel. This energy may be received at started planet or from space ship engine.

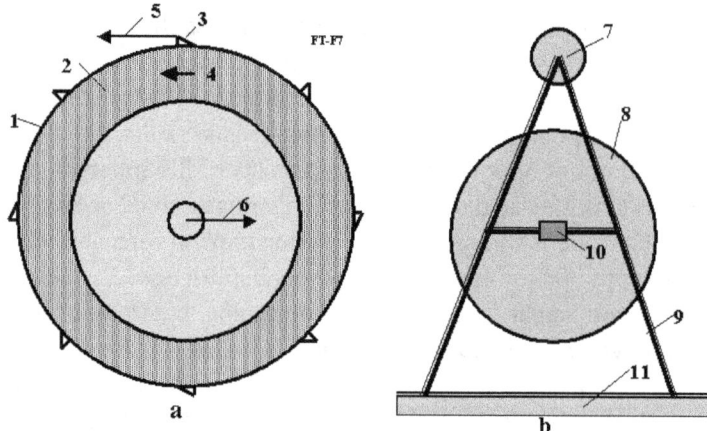

Fig. 7. Schema of new rocket and propulsion system. (**a**) Propulsion system from AB matter and storage energy. (**b**) Rocket with offered propulsion system.
Notations: 1 – cover (flywheel) from AB-matter; 2 – any work mass; 3 – nozzle with control of exit mass; 4 – direction of rotation; 5 – direction of exhaust mass; 6 – thrust; 7 – space ship; 8 – offered propulsion system; 9 – undercarriage; 10 – rotary mechanism; 11 – planet surface.

The rocket used the suggested engine is shown in fig, 7b. That has a cabin 7, the offered propulsion system 8, undercarriage 9 and rotary mechanism 10 for turning the ship in need position.

Let us to estimate the possibility of offered rocket. Notate, the relation of the exhaust mass to AM-matter cover mass of flywheel are taken $a = 10$, the safety (strength) factor $b = 4$. About 20% of space ship is payload and construction and 80% is the exhaust mass. Then exhaust speed of throw away mass and receiving speed by space ship are:

$$V = \sqrt{\frac{k}{ab}} = 2.12 \cdot 10^6 \ m/s, \quad m_s V_s = mV, \quad V_s = \frac{m}{m_s}V = \frac{0.8}{0.2} \cdot 2.12 \cdot 10^6 = 8.48 \cdot 10^6 \ m/s, \tag{25}$$

where V speed of exhausted mass, m/s; $k = \sigma/d = 1.9 \times 10^{14}$ (m/s)^2 is strength coefficient (16); m_s is final mass of rocket, kg; $V_s = 8480$ km/s is final speed of rocket, m/s; m is throw off mass, kg.

Let us to remind the escape speed of planets.

Table 1. Some data and escape speed from planets of Solar system.

Sun and planets	Distance from Sun 10^6 km	Gravity m/s^2	Escape speed km/s	Planets	Distance from Sun 10^6 km	Gravity m/s^2	Escape speed km/s
Sun	-	274	617.7	Jupiter	777.8	23.01	60.2
Mercury	57.9	3.72	4.15	Saturn	1326	9.14	36.2
Venus	108.1	8,69	10.25	Uranus	2868	9.67	21.4
Earth	149.5	9.78	11.19	Neptune	4494	15	23.4
Mars	227.8	3.72	5.09	Moon	0.384*	1.62	2.36

* From Earth.

The Table 1 allows estimating how many times the offered rocket can flight to other planets using one refueling (re-energy). These numbers are (with returning back to Earth): to Moon - 420 times, to Mars - 280 times, to Venus – 200 times, to Jupiter 96 times and out of Solar system 7 times.

3. Super-weapon.

Capability of an AB-Matter flywheel to spin up and ejection matter at huge speed (24)-(25) may be used as a long distance super-weapon.

4. Super-armor from conventional weapons.

The value (24) gives the need speed for break through (perforation) of a shield of AB-Matter. No weapon which can give this speed exists at the present time. Remain, the AB-Matter may be radiation impermeable. That means AB-Matter can protect from a nuclear bomb and laser weapon.

5. Simple thermonuclear reactor.

The AB-Matter film may be used as the wall of a simple thermonuclear reactor. The AB-Matter film allows a direct 100% hit by the accelerated nuclei to stationary nuclei located into film. You get a controlled nuclear reaction of cheap fuel. For example:

$$^1H + {}^1H \to {}^2H + e^+ + \upsilon + 0.42 \text{ MeV}, \quad {}^2H + {}^1H \to {}^3He + \gamma + 5.494 \text{ MeV}, \quad (26)$$
$$^2H + {}^2H \to {}^3H + {}^1H + 4.033 \text{ MeV}, \quad {}^3H + {}^1H \to {}^4He + \gamma + 16.632 \text{ MeV}. \quad (27)$$

Here e^+ is electron; υ is neutrino; γ is γ-quantum, photon (γ-radiation); ^1H = p is proton; ^2H = D is deuterium; ^3H = T is tritium; He is helium.

In conventional thermonuclear reactor the probability of a hit by the accelerated (or highly heated) nuclei to other nuclei is trifling. The accelerated particles, which run through ghostlike ATOMS and lose the energy, need therefore to be sent through to repeated collisions each of which loses energy until the one that hits and generates energy. The winner must pay for all the losers. That way we need big, very complex, and expensive high temperature conventional thermonuclear reactors. They are so nearly unbuildable because ordinary matter literally cannot take the reactions they are designed to contain, and therefore special tricks must be used to sidestep this, and the reactions are so improbable that again special tricks are required. Here, every shot is a hit and the material can endure every consequence of that hit. A good vacuum system and a means of getting power and isotopes in and out are the main problems, and by no means insuperable ones. Using the AB-Matter we can design a micro-thermonuclear AB reactor.

6. High efficiency rocket, jet and piston aviation engines.

The efficiency conventional jet and rocket engines are very limited by the temperature and safety limits of conventional matter (2000°K). If we will design the rotor blades (in jet engine), combustion chamber (in rocket and piston engines) from AB-Matter, we radically improve their capacities and simplify their construction (for example, no necessary cooling system!).

7. Hypersonic aircraft.

The friction and heat which attacks conventional materials for hypersonic aircraft limits their speed. Using the AB-Matter deletes this problem. Many designs for aerospace planes could capture oxygen in flight, saving hauling oxidizer and carrying fuel alone—enabling airliner type geometries and payloads since the weight of the oxidizer and the tanks needed to hold it, and the airframe strengths required escalate the design and cascade through it until conventional materials today cannot build a single stage to orbit or antipodes aerospace plane. But that would be quite possible with AB-Matter.

8. Increasing efficiency of a conventional aviation and transport vehicles.

AB-Matter does not experience friction. The air drag in aviation is produced up 90% by air friction on aircraft surface. Using AB-Matter will make jump in flight characteristics of aircraft and other transport vehicles (including sea ships and cars).

9. Improving capabilities of all machines.

Appearance new high strength and high temperature AB-Matter will produce jump, technology revolution in machine and power industries.

10. Computer and computer memory.

The AB-Matter film allows to write in 1 cm^2 $N = 1/(4 \times 10^{-26}) = 2.5 \times 10^{25}$ 1/cm^2 bits information. The current 45 nanometer technology allows to write only $N = 2.5 \times 10^{14}$ 1/cm^2 bit. That means the main chip and memory of computer based in AB-Matter film may be a billion times smaller and presumably thousands of times faster (based on the lesser distance signals must travel).

The reader can imagine useful application of AB-Matter in any field he is familiar with.

DISCUSSION

1. Pauli exclusion principle and Heisenberg Uncertainty Principle. General Question of Stability.

The reader may have questions about compatibility of the Pauli exclusion principle and Heisenberg Uncertainty Principle with AB-Matter. The uncertainty principle is

$$\Delta p \Delta x \geq h/2 . \qquad (28)$$

where $\Delta p = mV$ is momentum of particle, kg·m/s; m is mass particles, kg; V is speed particles, m/s; Δx is distance between particles, m; $h = 6.6262 \times 10^{-34}/2\pi$ is Planck's constant.

Pauli states that no two identical fermions may occupy the same quantum state *simultaneously*. A more rigorous statement of this principle is that, for two identical fermions, the total wave function is anti-symmetric. For electrons in a single atom, it states that no two electrons can have the same four quantum numbers, that is, if particles caracteristics n, l, and m_l are the same, m_s must be different such that the electrons have opposite spins.

The uncertainty principle gives a high uncertainty of Δp for nucleons and very high uncertainty for electrons into AB-Matter. But high density matter (of the same order as our suggested AB-Matter) EXISTS in the form of nuclei of conventional matter and on neutron stars. That is an important proof - this matter exists. Some may question its' ability to stay in a superdense state passively. Some may doubt its' stability free of the fierce gravitation of neutron stars (natural degenerate matter) or outside the confines of the nucleus. But there are reasons, not all stated here, to suppose that it might be so stable under normal conditions.

One proof was provided by Freeman Dyson [11] and Andrew Lenard in 1967, who considered the balance of attractive (electron-nuclear) and repulsive (electron-electron and nuclear-nuclear) forces and showed that ordinary matter would collapse and occupy a much smaller volume without the Pauli principle.

Certainly, however this very question of stability will be a key focus of any detailed probe into the possiblities of AB-Matter.

Readers usually ask: what is the connection (proton to proton) given a new element when, after 92 protons, the connection is unstable?

Answer: That depends entirely on the type of connection. If we conventionally join the carbon atom to another carbon atom, we then get the conventional piece of a coil. If we joint the carbon atom to another carbon atom by the indicated special methods, we then get the very strong single-wall nanotubes, graphene nano-ribbon (super-thin film), armchair, zigzag, chiral, Fullerite, torus, nanobud and other forms of nano-materials. That outcome becomes possible because the atomic force (van der Waals force, named for the Dutch physicist Johannes Diderik van der Waals, 1837-1923, etc.) is NON-SPHERICAL and active in the short (one molecule) distance. The nucleon nuclear force also is NON-SPHERICAL and they may also be active about the one nucleon diameter distance (Fig. 1). That means we may also produce with them the strings, tubes, films, nets and other geometrical constructions.

Note: The same idea may hypotheticaly be developed for **atto** (10^{-18} m), **zepto** (10^{-21} m), and **yocto** (10^{-24} m) technologies. It is known that nucleons consist of quarks. Unfortunately, we do not know yet about size, forces and interactions between quark and cannot therefore make predictions about atto or zepto-technology. One theory posits that the quark consists of preons. But we do not know anything about preons. The possibility alone must intrigue us for now. Where does it all end?

CONCLUSION

The thermal engineering and aerospace fields very much need new high properties materials better than any available today.

The author offers a design for a new form of nuclear matter from nucleons (neutrons, protons), electrons, and other nuclear particles. He shows that the new AB-Matter has most extraordinary properties (for example, (in varying circumstances) remarkable tensile strength, stiffness, hardness, critical temperature, superconductivity, super-transparency, ghostlike ability to pass through matter, zero friction, etc.), which are millions of times better than corresponded properties of conventional molecular matter. He shows how to design space system, aircraft, ships, transportation, thermonuclear reactors, and constructions, and so on from this new nuclear matter. These vehicles will have correspondingly amazing possibilities (invisibility, passing through any walls and amour, protection from nuclear bombs and any radiation, etc).

People may think this fantasy. But fifteen years ago most people and many scientists thought – nanotechnology is fantasy. Now many groups and industrial labs, even startups, spend hundreds of millions of dollars for development of nanotechnological-range products (precise chemistry, patterned atoms, catalysts, metamaterials, etc) and we have nanotubes (a new material which does not exist in Nature!) and other achievements beginning to come out of the pipeline in prospect. Nanotubes are stronger than steel by a hundred times—surely an amazement to a 19th Century observer if he could behold them.

Nanotechnology, in near term prospect, operates with objects (molecules and atoms) having the size in nanometer (10^{-9} m). The author here outlines perhaps more distant operations with objects (nuclei) having size in the femtometer range, (10^{-15} m, millions of times less smaller than the nanometer scale). The name of this new technology is femtotechnology.

Let us to explain the main thrust of this by analogy. Assume we live some thousands of years ago in a great river valley where there are no stones for building and only poor timber. In nature we notice that there are many types of clay (nuclei of atom—types of elemet). One man offers to people to make from clay bricks (AB-Matter) and build from these bricks a fantastic array of desirable structures too complex to make from naturally occuring mounds of mud. The bricks enable by increased precision and strength things impossible before. A new level of human civilization begins.

The author calls upon scientists and the technical community to to research and develop femtotechnology. He thinks we can reach in this field progress more quickly than in the further prospects of nanotechnology, because we have fewer (only 3) initial components (proton, neutron, electron) and interaction between them is well-known (3 main forces: strong, weak, electostatic). The different conventional atoms number about 100, most commone moleculs are tens thousands and interactions between them are very complex (e.g. Van der Waals force).

It may be however, that nano and femto technology enable each other as well, as tiny bits of AB-Matter would be marvellous tools for nanomechanical systems to wield to obtain effects unimaginable otherwise.

What time horizon might we face in this quest? The physicist Richard Feynman offeredhis idea to design artificial matter from atoms and molecules at an American Physical Society meeting at Caltech on December 29, 1959. But only in the last 15 years we have initial progress in nanotechnology. On the other hand progress is becoming swifter as more and better tools become common and as the technical community grows.

Now are in the position of trying to progress from the ancient 'telega' haywagon of rural Russia (in analogy, conventional matter composites) to a 'luxury sport coupe' (advanced tailored nanomaterials). The author suggests we have little to lose and literal worlds to gain by simultaneously researching how to leap from 'telega' to 'hypersonic space plane'. (Femotech materials and technologies, enabling all the wonders outlined here [12]).

REFERENCES

(The reader may find some of these articles at the author's web page http://Bolonkin.narod.ru/p65.htm , http://arxiv.org search "Bolonkin", and in the books "*Non-Rocket Space Launch and Flight*", Elsevier, London, 2006, 488 pages, "*New Concepts, Ideas, Innovations in Aerospace, Technology and Human Science*", NOVA, 2007, 502 pages and "*Macro-Projects: Environment and Technology*", NOVA 2008, 536 pages).

1. Bolonkin A.A., (1983a) Method of a Keeping of a Neutral Plasma and Installation for it. Russian Patent Application #3600272/25vv086993, 6 June 1983 (in Russian), Russian PTO.

2. Bolonkin A.A., (1983b) Method of transformation of Plasma Energy in Electric Current and Installation for it. Russian Patent Application #3647344/136681, 27 July 1983 (in Russian), Russian PTO.

3. Bolonkin A.A., Femtotechnology: Design of the Strongest Nuclear Matter with Fantastic Properties—AB-Matter. *American Journal of Engineering and Applied Sciences*. 2 (2), 2009, p.683-693. [on line] http://www.scipub.org/fulltext/ajeas/ajeas22683-693.pdf or http://sciprint.org, http://arxiv.org . (17 February 2009).

4. Bolonkin, A.A., *Non-Rocket Space Launch and Flight, Elsevier*, London, 2006, 488 pages. http://Bolonkin.narod.ru/p65.htm .

5. Bolonkin, A.A., "*New Concepts, Ideas, Innovations in Aerospace, Technology and Human Science*", NOVA, 2007, 502 pgs.   http://Bolonkin.narod.ru/p65.htm

6. Bolonkin A.A., and Cathcart R.B., *Macro-Projects in Environment and Technology*, NOVA, 2008, 500 pgs.   http://Bolonkin.narod.ru/p65.htm .

7. Encyclopedia of Physics. http://www.physicum.narod.ru (in Russian).

8, Tables of Physical values. Reference book, Editor I.K. Kikoin, Moscow, 1976, 1006 pgs. (in Russian).

9. AIP Physics Desk Reference, 3rd Edition, Springer, 2003. 888 pgs.

10. Dresselhaus, M.S., Carbon Nanotubes, Springer, 2000.

11. Wikipedia. Some background material in this article is gathered from Wikipedia under the Creative Commons license. http://wikipedia.org .

12. Bolonkin A.A., Converting of Matter to Nuclear Energy by AB-Generator. *American Journal of Engineering and Applied Sciences*. 2 (2), 2009, p.683-693. [on line] http://www.scipub.org/fulltext/ajeas/ajeas24683-693.pdf or http://sciprint.org.

13. Bolonkin A.A., Femtotechnology. Nuclear AB-Matter with Fantastic Properties, *American Journal of Enginering and Applied Sciences*. 2 (2), 2009, p.501-514. [On line]: http://www.scipub.org/fulltext/ajeas/ajeas22501-514.pdf, or http://www.podtime.net/sciprint/fm/uploads/files/1243447289Article%20Femtotechnology%20Design%20AB-Matter%20after%20Joseph%201%2028%2009.doc .

2008.

Chapter 4

AB-Needles: Stability and Production Super-Strong AB-matter (v.2)

Abstract

In works [1-3] author offered and considered possible super strong nuclear matter. In given work he continues to study the problem of a stability and production this matter. He shows the special artificial forms of nuclear AB-matter which make its stability and give the fantastic properties. For example, by the offered AB-needle you can pierce any body without any damage, support motionless satellite, reach the other planet, and research Earth's interior. These forms of nuclear matter are not in nature now, and nanotubes are also not in nature. The AB-matter is also not natural now, but researching and investigating their possibility, properties, stability and production are necessary for creating them.

Key words- Femtotechnology; FemtoTech; AB-matter; AB-needle; Stability AB-matter; Production of AB-matter;

INTRODUCTION

Brief History

Physicist Richard Feynman offered his idea to design artificial matter from atoms and molecules at an American Physical Society meeting at Caltech on December 29, 1959. If he was not well-known physicist, the audience laughed at him and drove away from the podium. All scientists accepted his proposal as joke. Typical question are: How can you see the molecule? How can you catch the molecule? How can you connect one molecule to other? How many hundreds of years you will create one milligram of matter? And hundreds of same questions having no answers may be asked. Any schoolboy has seen that Feynman proposal is full of fantasy which does not have relation to real technology. About 40 years the scientists had not found a way for implementation of this idea. But only in the last 15 years we have initial progress in nanotechnology. On the other hand, progress is becoming swifter as more and better tools become common and as the technical community grows. On 14 February 2009 the author offered the idea of design of new matter from protons, neutrons and electrons, made initial research and published the article about it [1]. These particles in million times are smaller than molecules. He researched and showed the new AB-matter will have the fantastic properties. That will be in millions times stronger than nanotubes and can keep the millions degrees of temperature. That may be invisible and permeable to ordinary matter. The many readers, who did not read carefully the author's article and who remembered from school course that the nucleus became unstable if number of protons is more than 92 or number of nucleons is more than 238, raised the cry that the AB-matter is impossible. They have not seen the **main difference** between conventional matter (conventional nucleus) and AB-matter. The conventional matter has nucleus which has a chaotic spherical LUMP (nucleus) of nucleons, the AB-matter is line from nucleons not having the lump. The author considered the AB-lines and shows in work [2] that lines is stable and has surprise property: one is a high rigid rod (needle), of which the compressed force does not depend on rod length! With this AB-rods (needles) you can support the Earth's satellite, reach the other planets, penetrate into the Earth interior and into any molecules of man without damage of its body.

Short Information About Offered Matter

In [1], it is shown the AB-matter may have forms Fig.1. The main forms are: "a"- single AB-string (AB-needle), "b"- AB-film (plate), and "d" is net. (see Fig. 5, Ch.3, p.33).

From AB-needles may be design the many other forms Fig.2 (see Fig. 6, Ch.3, p.34, taken from [1, Fig. 6]). That is net, cube, columns, tube and so on. For the reader's convenience we will repeat some of them (figs. 1-2).

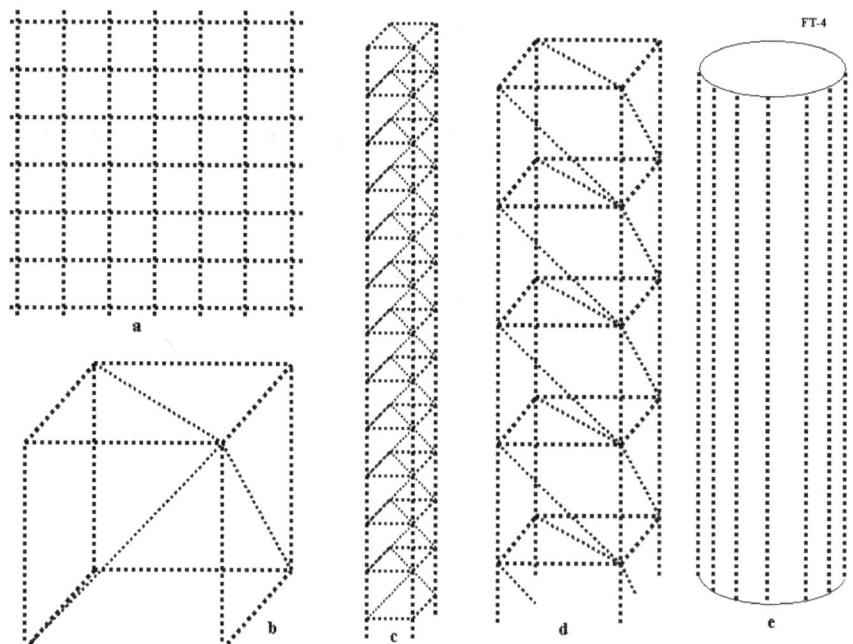

Fig. 1. Structures from nuclear AB-strings (AB-needles) (a) nuclear net (netting, gauze); (b) primary cube from matter strings; (c) primary column from nuclear strings; (d) large column where elements are made from primary columns; (e) tubes from matter strings (AB-needles) or matter columns.

AB-Matter

In conventional matter made of atoms and molecules the nucleons (protons, neutrons) are located in the nucleus, but the electrons rotate in orbits around nucleus in distance in millions times more than diameter of nucleus. Therefore, in essence, what we think of as solid matter contains a relatively 'gigantic' vacuum (free space) where the matter (nuclei) occupies only a very small part of the available space. Despite this unearthly emptiness, when you compress this (normal, non-degenerate) matter the electrons located in their orbits repel atom from atom and resist any great increase of the matter's density. Thus it feels solid to the touch.

The form of matter containing and subsuming all the atom's particles into the nucleus is named *degenerate matter*. Degenerate matter is found in white dwarfs, neutron stars and black holes. Conventionally this matter in such large astronomical objects has a high temperature (as independent particles) and a high gravity adding a forcing, confining pressure in a very massive celestial objects. In nature, degenerate matter exists stably (as a big lump) to our knowledge only in large astronomical masses (include their surface where gravitation pressure is zero) and into big nuclei of conventional matter.

Our purpose is to design artificial small masses of synthetic degenerate matter in form of an extremely thin strong thread (fiber, filament, string, needle), round bar (rod), tube, net (dense or non-dense weave and mesh size) which can exist at Earth-normal temperatures and pressures. Note that such stabilized special form matter in small amounts does not exist in nature as far as we know. Therefore author has named this matter AB-matter. Just as people now design the thousands variants of artificial materials (for example, plastics) from usual matter, we soon (historically speaking) shall create many artificial, designer materials by nanotechnology (for example, nanotubes: SWNTs (amchair, zigzag, ahiral), MWNTs (fullorite, torus, nanobut), nanoribbon (plate), grapheme, buckyballs (ball), fullerene). Sooner or later we may anticipate development of femtotechnology and create such AB-matter. Some possible forms of AB-matter are shown in Fig. 3

The main difference between the AB-matter and conventional matter is a strict order of location the proton and neutrons (for example: proton-neutron-proton-neutron) in line (string) or in the super thin (in one nucleon) plate (nuclear graphene). That gives the strong tensile stress (electrostatic repulse force) which does not allow the nucleons to mix in messy clump (ball). This force is less than a nuclear force if the AB-matter has a form where the most protons are located far from one another, where the nuclear force from the far protons is absent. That is in line, net and plate (Fig. 1a, b, d), but that may be absent in the solid beam, rod (Fig. 3c, d) if their cross-section area contains a lot of nucleons. The other problem: compensation of the positive charges is solved by rotating electrons around the AB string, rod, tube, net (grid) or an electron cloud near the plate [1] or the electron locates near nucleons.

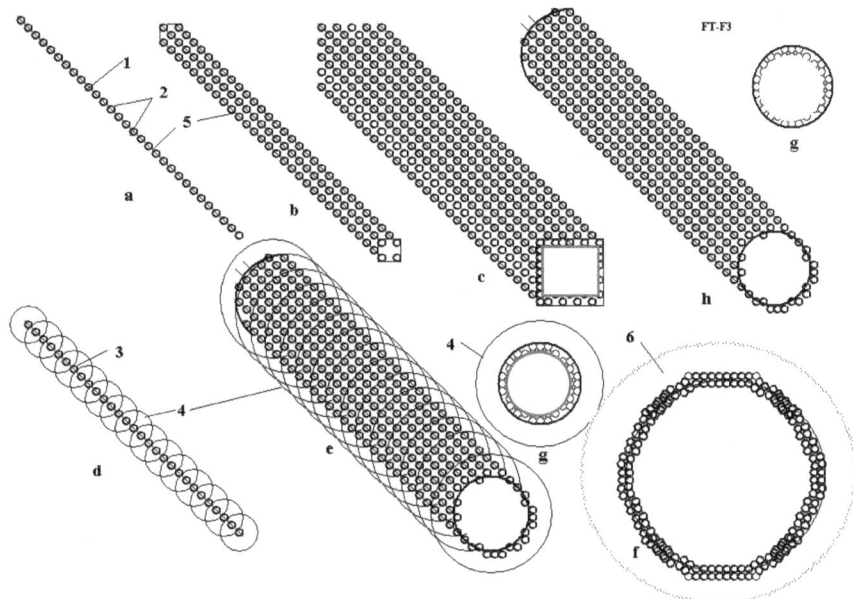

Fig. 2. Design of AB-matter from nucleons (neutrons, protons, etc.) and electrons (a) linear one string (monofilament) (fiber, whisker, filament, thread, needle); (b) ingot from four nuclear monofilaments; (c) multi-ingot from nuclear monofilament; (d) string made from protons and neutrons with electrons rotated around monofilament; (e) single wall femto tube (SWFT) fiber with rotated electrons; (f) cross-section of multi wall femto tube (MWFT) string; (g) cross-section of tube; (h) single wall femto tube (SWFT) string with electrons inserted into AB-matter. *Notations*: 1–nuclear string; 2-nucleons (neutrons, protons, etc.). 3-protons; 4-orbit of electrons; 5-nucleons; 6-cloud of electrons around tube.

Using the AB-matter

The simplest use of AB-matter is strengthening and reinforcing conventional material by AB-matter fiber. As is shown in the 'Computation' section [1], AB-matter fiber is stronger (has a gigantic ultimate tensile stress) than conventional material by a factor of millions of times, can endure millions degrees of temperature, and does not accept any attacking chemical reactions. We can insert (for example, by casting around the reinforcement) AB-matter fiber (or net) into steel, aluminum, plastic and the resultant matrix of conventional material increases in strength by thousands of times—if precautions are taken that the reinforcement stays put! Because of the extreme strength disparity design tricks must be used to assure that the fibers stay 'rooted'. The matrix form of conventional artificial fiber reinforcement is used widely in current technology. This increases the tensile stress resistance of the reinforced matrix matter by typically 2–4 times. Engineers dream about a nanotube reinforcement of conventional matrix materials which might increase the tensile stress by 10–20 times, but nanotubes are very expensive and researchers cannot decrease its cost to acceptable values yet despite years of effort. Another way is to use a construct of AB-matter as a continuous film or net (Fig. 2b, d) or as the AB-needles (Fig. 2) [3].

These forms of AB-matter have such miraculous properties as invisibility, superconductivity, zero friction, etc. The ultimate in camouflage, installations of a veritable invisible world can be built from certain forms of AB-matter with the possibility of being also interpenetable, literally allowing ghost-like passage through an apparently solid wall. Or the AB-matter net (of different construction) can be designed as an impenetrable wall that even hugely destructive weapons cannot penetrate.

The AB-matter film and net may be used for energy storage which can store up huge energy intensities and used also as rocket engines with gigantic impulse or weapon or absolute armor (see computation and application sections in [1-3]). Note that in the case of absolute armor, safeguards must be in place against buffering sudden accelerations; *g*-force shocks can kill even though nothing penetrates the armor!

The AB-matter net (which can be designed to be gas-impermeable) may be used for inflatable construction of such strength and lightness as to be able to suspend the weight of a city over a vast span the width of a sea. AB-matter may also be used for cubic or tower solid construction as it is shown in Fig. 3. Detailed computation of properties of the AB-matter is in [1-3]. Our purpose is to show that the curtain forms of AB-matter will be stable and may be produced.

ABOUT STABILITY OF THE NUCLEAR AB-MATTER

A. Short Information About Atom and Nuclei

Conventional matter consists of atoms and molecules. Molecules are collection of atoms. The atom contains a nucleus with proton(s) and usually neutrons (except for Hydrogen-1) and electrons revolve around this nucleus. Every particle may be characterized by parameters as mass, charge, spin, electric dipole, magnetic moment, etc. There are four forces active between particles: strong interaction, weak interaction, electromagnetic charge (Coulomb) force and gravitational force. The nuclear force dominates at distances up to $1.5 \div 2$ fm (femto, 1 fm = 10^{-15} m). They are hundreds of times more powerful than the charge (Coulomb) force and million-millions of times more than gravitational force. Charge (Coulomb) force is effective at distances over 2 fm. Gravitational force is significant near and into big masses (astronomical objects such as planets, stars, white dwarfs, neutron stars and black holes). Strong force is so overwhelmingly powerful that it forces together the positively charged protons, which would repel one from the other and fly apart without it. The strong force is key to the relationship between protons and neutrons. Electric force can keep electrons near nuclei. Scientists conventionally take into attention only the strong force when they consider the nuclear and near nuclear size range, and the other forces on that scale are negligible by comparison for most purposes.

Strong nuclear forces are anisotropic (non-spherical, force distribution not the same in all directions equally), which means that they depend on the relative orientation of the nucleus. The proton has a magnetic moment which produces the magnetic force. This force orients the proton in magnetic field and helps to keep the form of AB-matter.

Typical nuclear energy (force) is presented in Fig. 3 (or see Fig.1, Ch.3, p.32). The nuclear and electric forces can be attractive and repulsive. When it is positive the nuclear force repels the other atomic particles (protons, neutrons). When nuclear energy is negative, it attracts them up to a distance of about 2 fm. The value r_0 is usually taken as radius of nucleus.

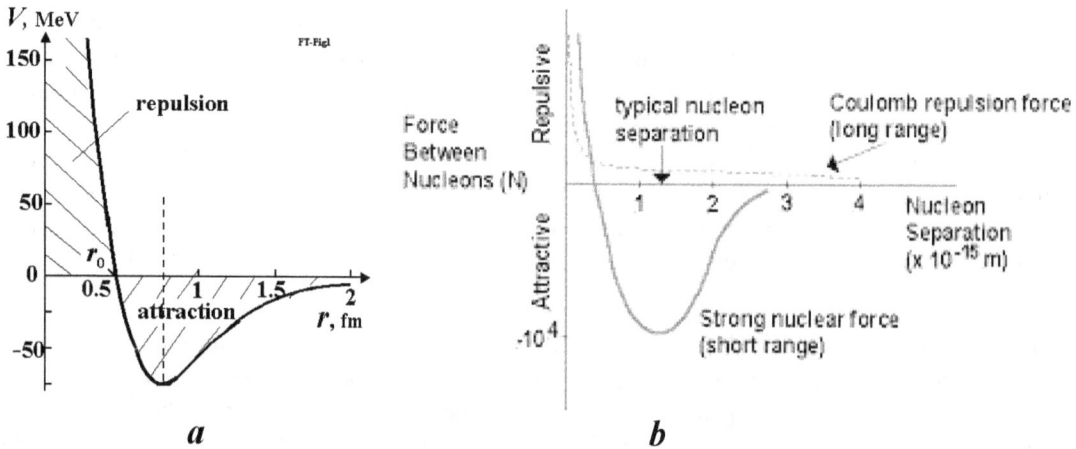

Fig. 3 Typical potencial (*a*) and nuclear force (*b*) of nucleus. When nucleon is at distance less than $1.8 \div 2.5$ fm, it is attracted to nucleus. When nucleon is very close, it is repulsed from nucleus [11].

B. Law (Necessary Conditions) of Stability the AB-matter

The necessary conditions (prerequisite law) of stability the AB-matter are as following:

1) The number of protons must be less approximately 90 into a local sphere of radius 3 fm in any point of AB-matter;

2) The number of nucleons must be less approximately 240 into a local sphere of radius 3 fm in any point of AB-matter;

3) The AB-matter contains minimum two protons.

4) Any neutron has minimum one contact with proton.

That law follows from relation between attractive nuclear and repulsive electrostatic forces into nucleus. The nuclear force is short distance force (2 fm), the electrostatic force is long distance force. When number of protons is more than 92 (in lamb), the repulsive electrostatic force may become more than nuclear force and electrostatic force may destroy the AB-matter. That law means the number of nucleons in any perpendicular cross-section area AB-matter design of Fig. 3 must be less than 37.

The results of computer simulation the potencial and forces in nucleos are shown in figs. 4a,b,c.

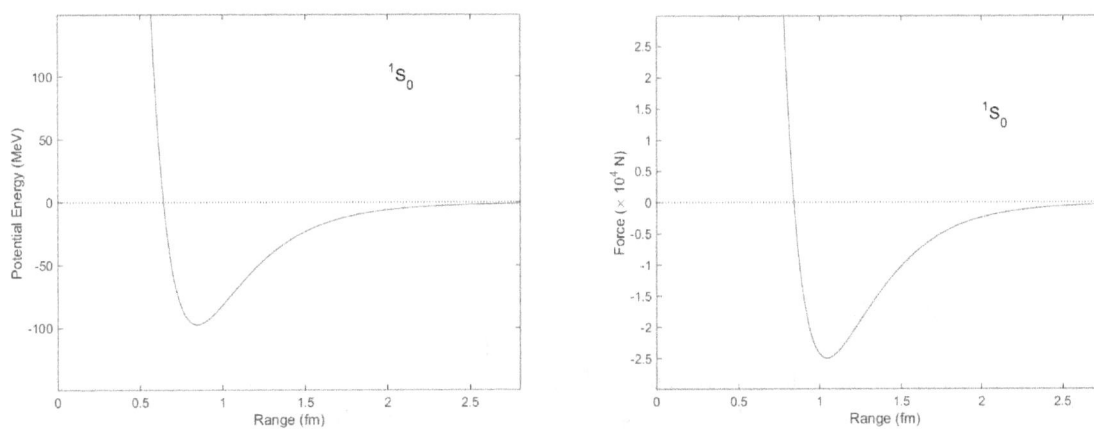

Fig. 4a (left). Corresponding potential energy (in units of MeV) of two nucleons as a function of distance as computed from the Reid potential. The potential well is a minimum at a distance of about 0.8 fm. With this potential nucleons can become bound with a negative "binding energy."

Fig.4b (right). Force (in units of 10,000 N) between two nucleons as a function of distance as computed from the Reid potential (1968).[1] The spins of the neutron and proton are aligned, and they are in the S angular momentum state. The attractive (negative) force has a maximum at a distance of about 1 fm with a force of about 25,000 N. Particles much closer than a distance of 0.8 fm experience a large repulsive (positive) force. Particles separated by a distance greater than 1 fm are still attracted (Yukawa potential), but the force falls as an exponential function of distance.

The press strong possibilities of the AB-matters are very large because AB-needles has the surprising property discovered by author – keep the huge press force in any length of AB-needle (transfer the pressure to any long distance). That properties are described in [3] and shortly in below.

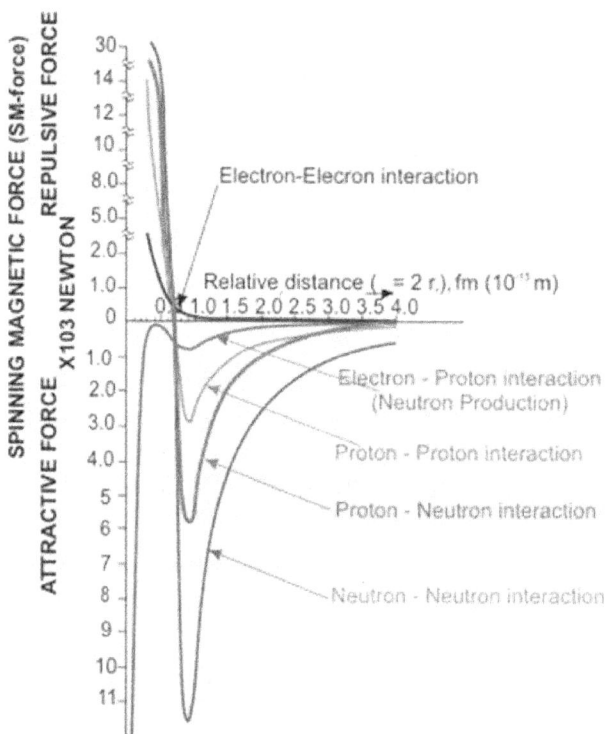

Fig.4c. Iteraction between electron, proton and neutron.
http://www.exmfpropulsions.com/New_Physics/SMFc.htm

AB-NEEDLES

The most important design of AB-matter is connection of nucleons in string (Fig. 5a, b, c). That may be only protons *pppp*.... (Fig. 5a), proton-neutron-proton-neutron-.... (*pnpn*....)(Fig. 5b), proton-neutron-neutron-proton-neutron-neutron-.... (*pnnpnn*....)(Fig. 5c). The ends of AB-string contains the protons. The electrostatic repulse force of these end protons is not balanced and creates the strong repulsive force 3 (Fig. 5c,d,e) which stretches the AB-string. That helps to keep the string form and other forms (plate, tube, beam, shaft, rod, etc.) of AB-matter presented in Figs. 3, 5. This is very important properties. This property does not have the conventional molecular matter, because the conventional matter contains the neutral molecules. The charges of ions in conventional matter locate far from one another and repulsive force is small. That property discovered by author gives the AB-string the amusing feature: an independence of the safety press stress from length of the nuclear string. Remand: the safety press force of long conventional matter strong depends on length of wire, beam, shaft, etc. According to the Euler's law the safety compressive force in the ordinary matter is inversely proportional to the square of the length of the rod. If the length of rod is more than the safety length, the construction losses the stability (one is bending). You cannot **push** the car a thread or thin wire having one km length. They bend. The AB thin string can pass the compressive force for any length of string. That is why it is named the AB-needle. AB-needle allows penetrating into any conventional matter, into the interior of Earth, planets, Sun. They allow making the interplanetary trips and investigations of planet from Earth.

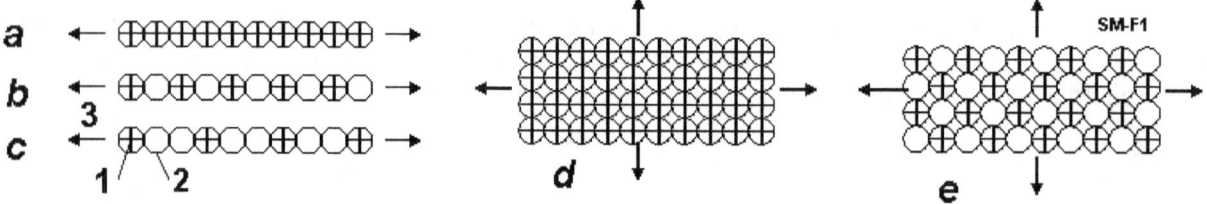

Fig. 5 Connection of nucleons in string (needle) (Fig. 2a, b, c) and film, plate (Fig.5d, e) and Coulomb (electrostatic, repulse) force. *Notations*: 1-protons, 2-neutrons, 3-repulse (Coulomb) force from protons.

Theory of AB-needles

Estimation of magnetic force the nucleons.

Proton and neutron have the magnetic momentum p_m. That means they are magnets. Magnets have North and Source poles and have ability connecting one to other in line. Let us to calculate this force and compare it with the nuclear and electrostatic forces.

The magnetic momentum p_m (J/T) is creating the circle currency I (fig. 6). It is

$$p_m = Is, \qquad (1)$$

where I is electric currency, A; $s = \pi r^2$ is a circle area of electric currency, m²; r is radius of circle, m.

Magnetic momentum of proton, neutron are known. The radius of the proton charge is also measured. In any case it is less than the radius of particles. From (1) we calculate the minimal electric currency of nucleon

$$I = p_m/s. \qquad (2)$$

The magnetic intensity H in point "A" located in circle axis OA (fig.5) is

$$H = \frac{Ir^2}{2(r^2+b^2)^{3/2}}. \quad \text{For} \quad \rho \gg r, \quad H \approx \frac{p_m}{2\pi b^3}. \qquad (3)$$

For $b = 0$ the magnetic intensity in circle center is

$$H_0 = I/2r \quad \text{or} \quad H_0 = p_m/2\pi r^3. \qquad (4)$$

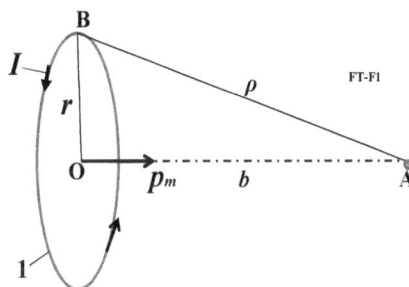

Fig.6. For theory of the magnetic intensity in an axis of the circle currency having the magnetic momentum p_m. *Notations*: A is point where we measure the magnetic intensity H; 1 – current circle is creating the electric charge: r is radius, I is electric currency, p_m is magnetic momentum; b is distance of line OA; ρ is distance AB.

The charge of proton is $e = +1.6 \cdot 10^{-19}$, C. That allows to calculate the number of charge revolutions n (1/s) and speed of charge V (m/s):

$$n = I/e, \quad V = 2\pi r n, \qquad (5)$$

where V must be less than the light speed $c = 3 \cdot 10^8$ m/s.

If we know the magnetic intensity, we can estimate the attractive force F (N) of opposed pole same (closed) magnet

$$F = \frac{\mu_0 H^2 s}{2} \quad \text{or} \quad F_0 = \frac{\mu_0 p_m^2}{8\pi r^4}, \qquad (6)$$

where $\mu_0 \approx 4\pi \cdot 10^{-7}$ is magnetic constant (permeability) (N/A²). F_0 is force (N) in center of the currence circle.

The computation shows: the magnetic forces have the significantly value. For example, the proton has $p_m = 1.41 \times 10^{-26}$ J/T, the radius $r = 0.8775 \times 10^{-15}$ m. From (6) we get $F_0 = 16.9$ N. This force is closed to the nuclear +

electric forces. The spherical distribution (as dot lines) of these forces along the radius are shown in fig. 7. As the result the attractive forces along the magnetic momentum axis is considerable more than in perpendicular axis. The force body has a form of rotated oval or ellipsoid. In result the protons and neutrons may be orientated in outer magnetic field and connected into AB-needle or filament as shown in fig.7d.

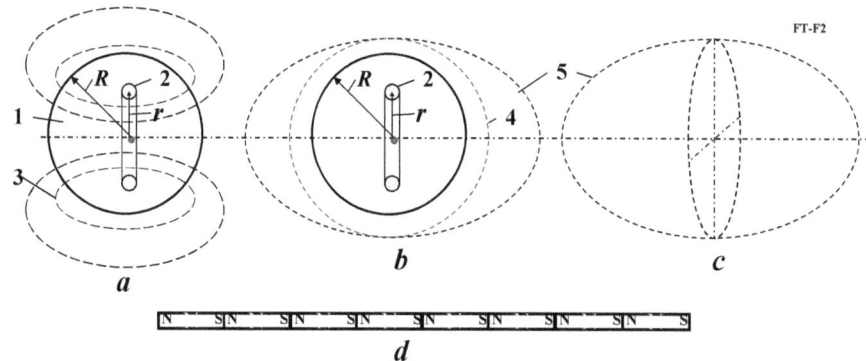

Fig.7. For computation of the magnetic force from the magnetic momentum of a nucleon (proton or neutron). *Notations*: *a* – nucleon: 1 – nucleon having radius *R*, 2 –electric currency ring having radius *r*, 3 – magnetic lines from the magnetic momentum. *b* – magnitudes of forces near nucleon: 4 – nuclear and electrostatic forces, 5 - magnetic force. Magnitudes equal a distance from the center to the dot line. *c* – form of forces in isometric view. Form is closed to a rotated oval or rotated ellipsoid. *d* – is AB needle from nucleons connected in line as magnets.

The magnetic forces of nucleons may be significantly increased when they connected one to other (fig.8a). In this case they work as solenoid having the summary electric currency and an inverse linear relationship from a length of solenoid *d* (not inverse relationship of the third order as in (4)-(3)).

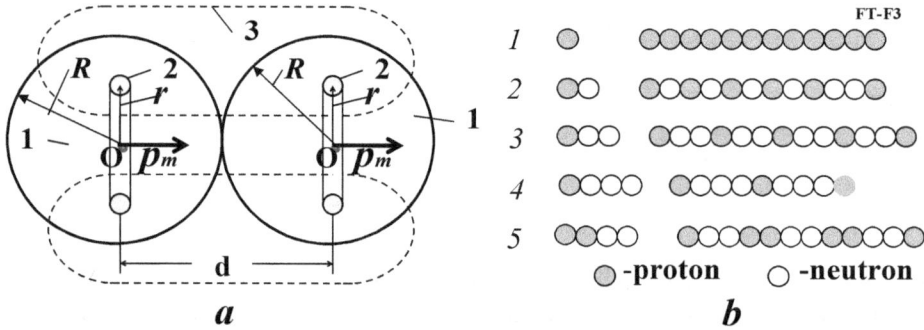

Fig.8. Left: Two connected nucleus work as solenoid. Their magnetic force is increases. 3 is common magnetic lines. Right: AB-needles: *Notations:* *1* – AB-needle from protons (*pppp*…); *2* – AB-needle from one proton + one neutron (*pnpn*…); *3* – AB-needle from one proton + two neutrons (*pnnpnn*…); *4* – AB-needle from one proton + three neutron (pnnnpnnn…), *5* - AB-needle from two protons + two neutrons (*ppnnppnn*…)(ends have one proton).

The magnetic intensity (*H*) and force between two nucleons **as solenoid** may be computed the equations:

$$H = \frac{I_1 + I_2}{d}, \quad F_d = \frac{\mu_0 H^2 s}{2},$$
(7)

where I_1 is circle electric currency in the first nucleon (A), I_2 is circle electric currency in the second nucleon (A); *d* is distance between circles (m), $d \approx 2R$; F_d is force, N; $s = \pi r^2$ is a circle area of the electric currency, m². If solenoid row have *n* same nucleons, the summary magnetic intensity of long solenoid changes as:

$$H = \frac{I}{d}\frac{n+1}{n}, \quad F_d = \frac{\mu_0 H^2 s}{2}, \quad \lim \frac{n+1}{n} \to 1 \quad \text{for} \quad n \to \infty, \quad n \geq 1,$$
(8)

The AB-needles can have the different structures. Some of them are sown in fig. 8b. For example: The first AB-needle (fig.8b-*1*) contains only protons (*pppp*…). The second (fig. 8b-*2*) - contains the proton-neutron (pnpn…). The third (fig. 8b-*3*) - contains proton and two neutrons (pnnpnn…). The fourth (fig.8b-*4*)– contains proton + three neutrons (pnnnpnnn…). The fifth (fig.8b – *5*) - contains two protons + two neutrons (ppnnppnn…). I think for stability AB-needle (include many row needles about every neutrons must be proton because some theory propose the proton and neutron are exchanging the charge.

The repulse electrostatic force between two protons is

$$F_e = k\frac{e^2}{d^2}, \qquad (9)$$

where $k = 9 \times 10^9$ N·m²/C² is electrostatic constant; d is distance between two protons, m. If row (thread) contains only protons the maximal summary electrostatic force is

$$F_e = k\frac{e^2}{d^2}\sum_1^N \frac{1}{n^2}, \quad \lim\sum_1^N \frac{1}{n^2} = \frac{\pi^2}{6} \quad \text{for} \quad N = \infty, \quad n \geq 1, \qquad (10)$$

The average nuclear force F_N (N) equals approximately

$$F_N = E/l,$$

where E is avarage nuclear connection (binding) energy (J), l is average distance of the active force (m).

For the hydrogen nucleus ²H (proton + neutron) the nuclear binding energy equals 1 MeV [12, p.547]. That means $E = 1{,}6 \times 10^{-13}$ J. The active distance of the nuclear force is about $l = 10^{-15}$ m. Substitute it into equation (11) we get the average nuclear force

$$F_N = 160 \text{ N}.$$

Numerical estimation.

1. Let us estimate the forces between **proton - proton**. It is known the radius R (m) of nucleon approximately equals

$$R \approx 1.2 \times 10^{-15} \sqrt[3]{A}, \qquad (11)$$

where A is number of nucleons into nucleus. In our case $A = 1$ and $R \approx 1.2 \times 10^{-15}$ m for all nucleons (proton and neutron). The radius of positive charge into proton is known and equal $r = 0.8775 \times 10^{-15}$ m. The magnetic momentum of proton is $p_m = 1.41 \times 10^{-26}$ J/T.

Using the equation above we can compute magnetic force and next parameters of **solenoid** from couple neutrons:

$$s = \pi r^2 = 3.14 \times (0.8775 \times 10^{-15})^2 = 2.42 \times 10^{-30} \; m^2, \qquad (12)$$

$$I = \frac{p_m}{s} = \frac{1{,}41 \cdot 10^{-26}}{2.42 \cdot 10^{-30}} = 5.83 \cdot 10^3 \; A, \quad n = \frac{I}{e} = \frac{5.83 \cdot 10^3}{1.6 \cdot 10^{-19}} = 3.64 \cdot 10^{22} \; \frac{1}{s}, \qquad (13)$$

$$V = 2\pi r n = 2 \cdot 3.14 \times (0.8775 \times 10^{-15}) \cdot 3.64 \cdot 10^{22} = 2 \times 10^8 \; m^2. \qquad (14)$$

For solenoid having two protons we get:

$$H = \frac{2I}{d} = \frac{2 \cdot 5.83 \cdot 10^3}{2 \cdot 1.2 \cdot 10^{-15}} = 4.86 \times 10^{18}, \; A/m,$$

$$F_n = \frac{\mu_0 H^2}{2} s = \frac{4\pi \cdot 10^{-7}(4.86 \times 10^{18})^2}{2} 2.42 \cdot 10^{30} = 35.9 \; N, \qquad (15)$$

$$F_e = -k\frac{e^2}{d^2} = -9 \cdot 10^9 \frac{(1.6 \cdot 10^{-19})^2}{(2.4 \cdot 10^{-15})^2} = -40 \; N.$$

Note: the attractive nuclear force F_N =160 N (12) is significantly more than repulsive electrostatic force F_e = - 40N.

2. Let us estimate the solenoid magnetic force in couple **proton + neutron**. The magnetic momentum of neutron is $p_m = -0.966\times 10^{-26}$ J/T. Sign minus means the vector of magnetic moment and vector of mechanical momentum (spin) is opposed. It is not important for us because we can turn the vector of the magnetic force in a need direction by an outer magnetic field.

Let us take the radius of neutron equals the radius of proton $R = 1.2\times 10^{-15}$ m and radius of charge into neutron equal to radius of charge of proton $r = 0.8775\times 10^{-15}$ m.

Using the equation above we can compute the next values and force in neutron:

$$s = \pi r^2 = 3.14\times(0.8775\times 10^{-15})^2 = 2.42\times 10^{-30}\ m^2, \tag{16}$$

$$I_1 = \frac{p_m}{s} = \frac{0.966\cdot 10^{-26}}{2.42\cdot 10^{-30}} = 4\cdot 10^3\ A, \quad n = \frac{I}{e} = \frac{4\cdot 10^2}{1.6\cdot 10^{-19}} = 2.5\cdot 10^{22}\ \frac{1}{s}, \tag{17}$$

$$V = 2\pi r n = 2\cdot 3.14\times(0.8775\times 10^{-15})\cdot 2.5\cdot 10^{22} = 1.21\times 10^8\ m/s. \tag{18}$$

For solenoid having two protons we get:

$$H = \frac{I + I_1}{d} = \frac{5.83\cdot 10^3 + 4\cdot 10^3}{2\cdot 1.2\cdot 10^{-15}} = 4.1\times 10^{18},$$

$$F_n = \frac{\mu_0 H^2}{2}s = \frac{4\pi\cdot 10^{-7}(4.1\cdot 10^{18})^2}{2}2.42\cdot 10^{30} = 25.5\ N, \tag{19}$$

The attractive nuclear force $F_N = 160$ N (12).

3. Let us estimate the solenoid magnetic force in couple **neutron + neutron**. The magnetic momentum of neutron is $p_m = -0.966\times 10^{-26}$ J/T. Sign minus means the vector of magnetic moment and vector of mechanical momentum (spin). Let us take the radius of neutron equals the radius of proton $R = 1.2\times 10^{-15}$ m and radius of charge into neutron equal to radius of charge of proton $r = 0.8775\times 10^{-15}$ m.

Using the equation above we can compute the next values and force in neutron:

$$s = \pi r^2 = 3.14\times(0.8775\times 10^{-15})^2 = 2.42\times 10^{-30}\ m^2, \tag{20}$$

$$I = \frac{p_m}{s} = \frac{0.966\cdot 10^{-26}}{2.42\cdot 10^{-30}} = 4\cdot 10^3\ A, \quad n = \frac{I}{e} = \frac{4\cdot 10^2}{1.6\cdot 10^{-19}} = 2.5\cdot 10^{22}\ \frac{1}{s}, \tag{21}$$

$$V = 2\pi r n = 2\cdot 3.14\times(0.8775\times 10^{-15})\cdot 2.5\cdot 10^{22} = 1.21\times 10^8\ m/s. \tag{22}$$

For solenoid having two protons we get:

$$H = \frac{2I_1}{d} = \frac{2\cdot 4\cdot 10^3}{2\cdot 1.2\cdot 10^{-15}} = 3.33\times 10^{18},$$

$$F_n = \frac{\mu_0 H^2}{2}s = \frac{4\pi 10^{-7}(3.33\cdot 10^{18})^2}{2}2.42\cdot 10^{30} = 16.9\ N, \tag{23}$$

The attractive nuclear force $F_N = 160$ N.

This computations show: the magnetic forces of protons and neutrons allow to design from them the long AB-treads if we connect them by the corresponding magnetic poles. We can make their need orientation by the outer magnetic field. If AB-thread contains the correct located protons one became the springy AB-needle.

IV. SOME APPLICATION of AB-needles

Some constructions from AB-string are shown in Fig.9.

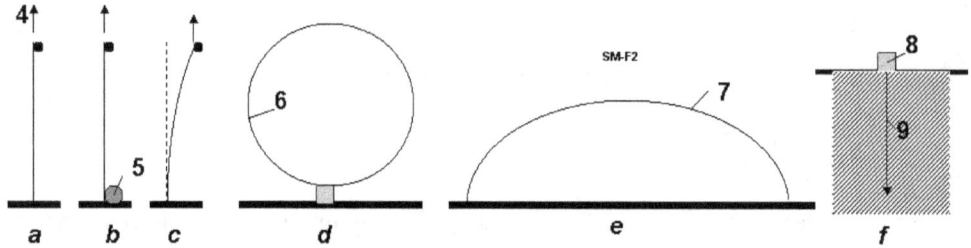

Fig. 9. Some construction from AB-string. *Notations*: *a* – vertical string (AB-needle). The big lift (support) force 4 does not depend from length; *b* – lifting the load to any altitude. 5 - spool of AB-string; *c* – stability of AB-string; *d* – ring 6 from AB-string; *e* – bridge (long arm) from AB-string; *f* – research of the Earth crust interior: 8 - installation (spool of AB-needle), 9 – AB-needle (string, cable).

AB-needles may be illustrated by a children long inflatable air-balloon (Fig. 10a). This press force also does not depend on length of balloon. The force is transferred by compressed air. This idea was used by author in designing the inflatable space tower [5].

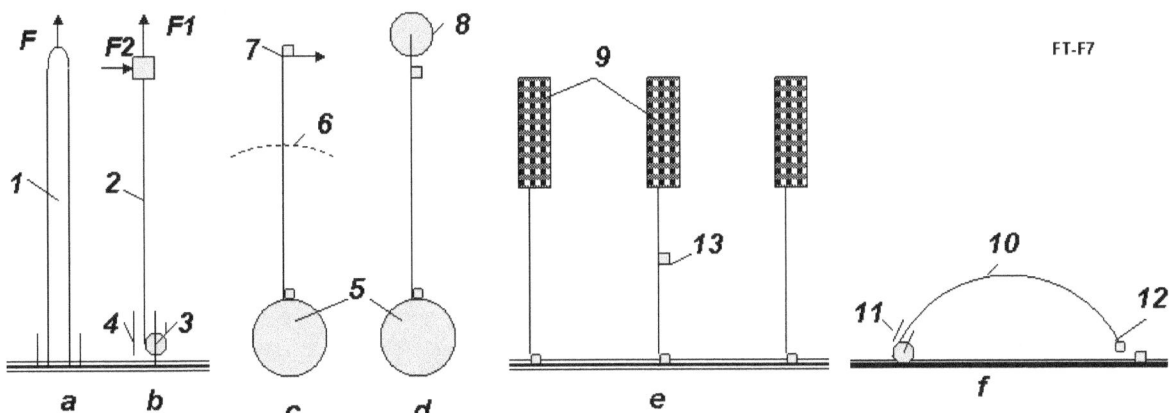

Fig. 10. Applications of AB-needles. *Notations*: *a* – conventional children inflatable long tube illustrated the capability to accept the pressure in end of tube (*F* – force); *b* – illustration of AB-needle to lift the load, accepts the vertical and horizontal forces (*F*1, *F*2 = 0.5*F*1); *c* – AB-needles as the over GSO Space Elevator; *d* – AB-needles as space ship and the investigator of the planet interior (for example Moon); *e* – the building suspended at high altitude by AB – needles, *f* – the investigation of interior of building, men, etc. by AB-needles. 1- conventional children inflatable long tube (air balloon); 2 – AB-needles; 3 – reel of AB-needles; 4 – the guides of AB-needles; 5 – Earth; 6 – Geosynchronous orbit; 7 – space ship; 9 - building; 10 – AB-needle; 11 - the guides of AB-needles; 12 – devices (TV-camera, capture grid, weapon, etc.); 13 – elevator.

The tension F_p activates along all lengths of AB-needle and does not allow to curl the AB-string into the lamb – conventional nucleus. This tension works when there are no other closed protons with a side of the string. When AB-needle is created, the outside protons cannot joint to AB-needle because the protons repel each other. The proton and neutron have the magnetic dipole moments. Magnetic dipole moment of proton equals $+1.41 \cdot 10^{-26}$ J/T, and magnetic dipole moment of neutron equals $-0.966 \cdot 10^{-26}$ J/T. They are small magnets having magnetic force some newtons. That also allows creating the stable AB-needles, to arrange them in a certain position and order.

The AB-needle can also keep the maximal side force $F2 \approx 0.5F1$ (Fig. 10b). That allows accelerating anybody (for example space ship) in side direction, to produce an elastic design (for example, air bridge, storage of mechanical energy, long arm (hand), etc.). AB-matter designs do not have the drawbacks of the ordinary matter as fatigue, residual strain and the susceptibility to the external environment.

One meter of AB-needle has line having $n = 5.7 \cdot 10^{14}$ nucleons with mass $m = 1.67 \cdot 10^{-27}$ kg. Total mass of one meter AB-needle equals only 10^{-12} kg/m.

$$M_1 = nm = 5.7 \cdot 10^{14} \times 1.67 \cdot 10^{-27} = 10^{-12} \text{ kg/m}.$$

One million kilometers of AB-needle weights only 10^{-3} kg/Mm. For transferring the large force we can take the thin cable from AB-needles.

A. **Summary**

Four above necessary conditions, repulsive force of protons and magnetic force of nucleons can make the stability of AB-matter.

PRODUCTION OF AB-NEEDLES

The charged particles interact with electric and magnetic fields. The magnetic moment interacts with magnetic field. That allows designing the technologies for production of artificial AB-matter. Some offered technologies were described in [1]. Here the author offers some new technologies.

The possible particles are shown in Table 1.

TABLE 1. CHARGE, IMPULSE AND MAGNETIC MOMENTS OF SOME NUCLEUS

Z	Nucleus (particles)	Charge $+e=1.6\cdot 10^{-19}$ C	Mass number	Impulse moment, $\hbar$	Magnetic* moment, μ_N
0	n	0	1	1/2	-1.9125
1	p	1	1	1/2	2.7828
1	^{2}H = D	1	2	1	0.8565
2	^{3}He	2	3	1/2	-2.121
2	^{4}He	2	4	0	0
3	^{6}Li	3	6	1	0.821
3	^{7}Li	3	7	3/2	3.2332

*Nuclear magnetron $\mu_N = 5.051\cdot 10^{-27}$ J/T. Sign "-"shows: magnetic moment is opposite the impulse moment.

A. Notes About Possible Form AB-needles

The possible form of AB-needles is shown in Fig. 11.

The first form marked 1 (pppp…) contains only line of protons. This form is cheapest and has maximum pressure strength. But it is unknown whether this form is possible or not. It is known the single hydrogen and single proton are stable. In other side the fusion of two single hydrogen nuclei ^{1}H (protons) produces deuterium ^{2}H= D (*pn*) releasing a positron and a neutrino as one proton changes into a neutron:

$$^1H + {}^1H \rightarrow {}^2H + \underline{e^+} + \underline{\nu_e} + 0.42 \text{ MeV} \qquad (24)$$

The fusion reaction released in this step produces energy up to 0.42 MeV. The most of this energy is taken away by neutrino.

The positron immediately annihilates with an electron, and their mass energy is carried off by two gamma ray photos:

$$e^+ + e^- \rightarrow 2\gamma + 1.02 \text{ MeV} . \qquad (25)$$

But most nucleuses have a lot of protons and they do not rely on the reaction (24). The AB-needle also has a lot of protons. If reaction (24) is released, the form 1 transfers in form 2 (Fig. 11) and the process produces a lot of nuclear energy. The ionized conventional hydrogen ^{1}H may be used for production of AB-matter. I remain: the Universe is composed of about 80% hydrogen. As a result we will have the AB-needle in form *npnp…* .

The second form of AB-needle is *pnpn…* marked 2 (Fig. 11). This form may be produced directly from deuterium *D* oriented by magnetic field along axis of AB-needles. The third form of the double AB-needles marked 3 (Fig. 10) may be also produced directly from deuterium *D* oriented by magnetic field perpendicular of axis of AB-needles. The forth form of four-needles marked 4 (Fig. 11) may be produced directly from helium ^{4}He oriented by magnetic field perpendicular of axis of AB-needles.

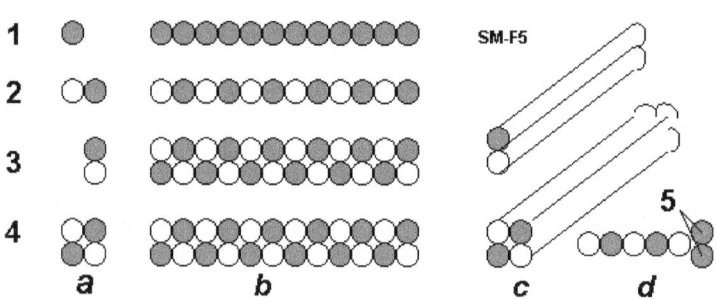

Fig.11 Types of AB-needles. *Notations*: *a* – Nucleus: black is *p*, white is *n*; *b* – AB-needles (side view); c – AB-needles in isometrical view; *d* – increasing the internal tensile stress by the double protons (5) located in the end of single AB-needle from protons (for increasing the tensile stress); 1- protons (*p*). Single AB-needles from proton; 2 – deuterium ^{2}H = D (*pn*). Single AB-needles from deuterium; 3 - deuterium ^{2}H (*pn*). Double AB-needles from deuterium; 4 – helium ^{4}He. 4 – square AB-needles from helium. 5 – double protons in end of single AB-needle.

B. Installations for Production AB-needles

1) The First Method: Toroid Method:

One of installation for production of AB-needles is shown in Fig. 12. The installation has a vacuum topoid 1 and particles gun 4 which injects charged particles into toroid. The perpendicular (to fig.) magnetic lines 2 penetrate the toroid. As a result the charged particles 3 move in circles inside the toroid. This electric current of particles produces the magnetic field 5 (pinch-affect). This field pulls the particles in a cord and helps to keep them into the toroid ring.

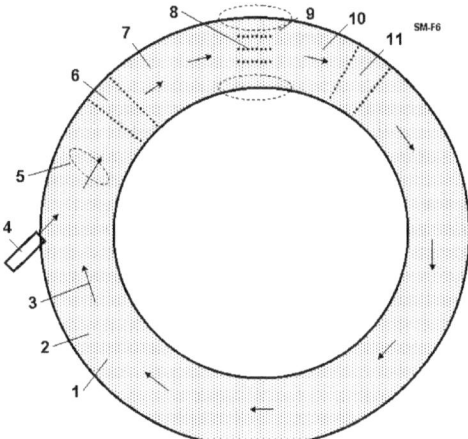

Fig. 12 Toroid producer of AB-needles (AB-matter). *Notations*: 1 – vacuum toroid; 2 – perpendicular (to sketch) magnetic lines; 3 – particles; 4 - particles gun; 5 – round magnetic lines from motion charged particles; 6 – electric accelerator; 7 – electric focuser; 8 – AB – needles; 9 – magnetic field keeping the AB-needles; 10 – electric focuser; 11 – electric accelerator.

The producing AB-needles 8 locate inside the toroid ring and are kept by special local magnetic field 9 in position along the circle axis of the toroid ring. That means the moving particles can connect to AB-needles only to end nucleus when they collide the forward end of AB-needle and their energy is sufficient to overcome the Coulomb repulsion. The toroid ring has the accelerators 6, 11 and focusers 7, 10 of particles. Their electric fields collect the scattered charged particles back to toroid axis.

Probability of hitting in the front end of the AB-needles is small. But the charged particles rotate into toroid a lot (millions) of times and join to end of AB-needles. Note they can connect only to end of AB-needle. Their perpendicular speed to the toroid circle axis is not enough to overcome the nuclear repulsion force.

Author wrote only the principal scheme (schematic diagram) of the AB-needle producing. The developing of this method may request a big research and work.

2) The Second Method: Method Particles Traps:

That is shown in Fig. 13. That is closed to method described in [1-3]. Feature is the net of traps 8 (Fig. 13a and 13b). They catch the particles and direct them to end of creating AB-needles. Advantage is high efficiency of production AB-matter (every charged particle will be used, small of energy consumption). Lack is the request of a special form of AB-matter (see 8 in Fig. 13b). That method may be useful when we have enough AB-matter.

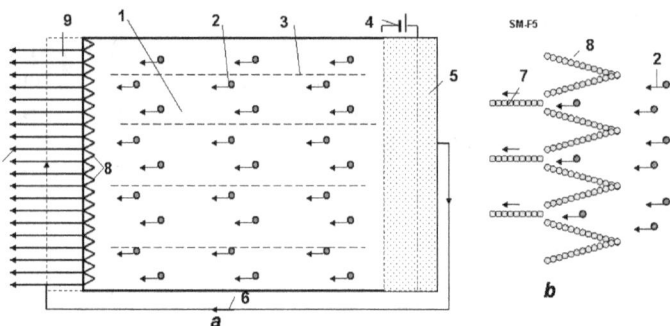

Fig. 13. Method particle traps for production of the AB-needles. *Notations*: a – device; b – particle traps; 1 - vacuum cell; 2- charged particles; 3 – magnetic lines; 4 – electric issue for the acceleration nets; 5 - plasma from particles; 6 – flow of electrons; 7 – AB-needles; 8 – trap made from AB-matter for the charged particles (p, ^{2}H, ^{4}He, etc.); 9 – cell for cover the AB-needles by electrons.

3) The Third Method: Method Standing Waves:

The current special mirrors [4, Ch.12] and lasers allow to create the net of electromagnetic traps for AB-matter producer (Fig. 14) from the monochromatic polarized electromagnetic standing waves (Fig. 14a, b). That net may partially change the net of AB-matter traps of the Fig. 14b and increase the efficiency. This method may be useful for AB-matter producer in [1-3].

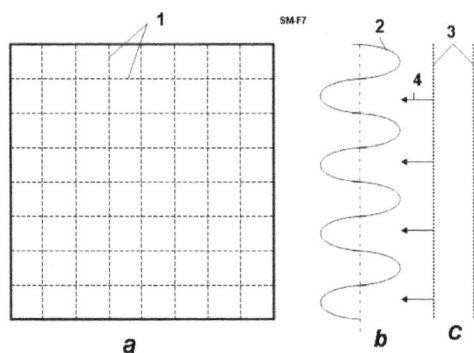

Fig. 14. Net of electromagnetic traps for AB-matter producer. *Notations*: a – forward view; b – the monochromatic polarized electromagnetic standing waves (electrostatic part, side view); c – particles storage and accelerator; 1 – net from the perpendicular monochromatic polarized electromagnetic standing waves; 2 - the electromagnetic monochromatic polarized standing wave; 3 – electric accelerator of particles; 4 – particles.

The threads from AB-matter are stronger by millions of times than normal materials. They can be inserted as reinforcements, into conventional materials, which serve as a matrix, and are thus strengthened by thousands of times (see computation section in [1]).

The offered AB-producers can be used for producing the new NANO-matters. Now the scientist offers to produce nano-matters by nano-robots. I think that is a very difficult way. The nano-robot must have the devices for searching, recognizing, catching the flying molecules, deliver them in given place, and connect to other selected molecules. That means the nano-robot must have a million molecules. It is difficult to get an elephant to catch the flies and glue them from the device. This productivity will be very low. The production of AB-matter may be easy.

Also we can ionize the molecules (create the charged particles!) and apply the modified offered methods for design and production of the nano-matters.

VI. DISCUSSION

The humanity will make a gigantic jump in technology when one will produce AB-matter. We consider unconventional application of AB-matter.

A. Super Micro-World from AB-Matter: An Amusing Thought-Experiment

AB-matter may have $10^{15} \div 10^{43}$ times more particles in a given volume than a single atom. A human being, man is made from conventional matter, contains about 5×10^{26} molecules. That means that 'femto-beings' of equal complexity from AB-matter (having same number of components) could be located in the volume of one microbe having size $10\ \mu = 10^{-5}$ m. It is difficult to make the nano-robot (one is large for Nano World). But the smart small femto-robot is suitable for Nano World. In future the people could make the artificial intelligent super micro F-beings which can withstand a huge temperature, acceleration of electric field, travel to other stars, other galactic, live in stars and travel through black holes to other universes and times.

B. Stability of AB-matter

Readers usually ask: the connection (proton to proton) gives a new element when, after 92 protons, this element is unstable?

Answer: That depends entirely on the type of connection. If we conventionally join the carbon atom to another carbon atom a lot of times, we then get the conventional piece of a coal. If we join the carbon atom to another carbon atom by the indicated special forms, we then get the very strong single-wall nanotubes, graphene nano-ribbon (super-thin film), armchair, zigzag, chiral, fullerite, torus, nanobud and other forms of nano-materials. That outcome becomes possible because the atomic force (van der Waals force, named for the Dutch physicist Johannes Diderik van der Waals, 1837-1923, etc.) is non-spherical and active in the short (one molecule) distance. The nucleon nuclear force is also non-spherical and they may also be active about the one nucleon diameter distance (Fig. 1). Moreover, the nucleus are tensile electrostatic force which allows designing the long linear structures. Moreover, the proton and neutron are the small magnets. The magnets (and nucleus) connect one to other specific side. That means we may also produce with them the strings, tubes, films, nets and other geometrical constructions.

The further studies are shown that AB-matter will be stable if:

1) The any sphere having radius $R \approx 6 \times 10^{-15}$ m in any point of structure Figs. 1- 4 must contain no more than 238 nucleons (about 92 of them must be protons). That means any perpendicular cross-section area of the solid rod, beam and so on of AB-structure (for example figs. 1b,c,g) must contain no more than about 36 nucleons in any circle with $R \approx 6 \times 10^{-15}$ m.

2) AB-matter must contain the proton in a certain order because the electrostatic repel forces of them give the stability of the given structure.

3) The magnetic force of protons and neutrons also allows giving the different forms of AB-matter.

VII. CONCLUSION

The author offers a design for a new form of nuclear matter from nucleons (neutrons, protons), electrons, and other nuclear particles. He also suggests the necessary conditions of stability of AB-matter. He shows that the new AB-matter has most extraordinary properties (for example, (in varying circumstances) remarkable tensile strength, stiffness, hardness, critical temperature, superconductivity, super-transparency, ghostlike ability to pass through matter, zero friction, etc.), which are millions of times better than corresponded properties of conventional molecular matter. He shows (in [2]) how to design aircraft, ships, transportation, thermonuclear reactors, and constructions, and so on from this new nuclear matter. These vehicles will have correspondingly amazing possibilities (invisibility, passing through any wall and amour, protection from nuclear bombs and any radiation, etc).

People may think this is fantasy. But fifteen years ago most people and many scientists thought nanotechnology is fantasy. Now many groups and industrial labs, even startups, spend hundreds of millions of dollars for development of nanotechnological-range products (precise chemistry, patterned atoms, catalysts, metamaterials, etc) and we have nanotubes (a new material which does not exist in Nature!) and other achievements beginning to come out of the pipeline in prospect. Nanotubes are stronger than steel by a ten times—surely an amazement to a 19th century observer if he could behold them.

Nanotechnology, in near term prospect, operates with objects (molecules and atoms) having the size in nanometer (10^{-9} m). The author here outlines perhaps more distant operations with objects (nuclei) having size in the femtometer range, (10^{-15} m, millions of times smaller than the nanometer scale). The name of this new technology is femtotechnology.

I want to explain the main thrust of this by analogy. Assume some thousands of years ago we live in a great river valley where there are no stones for building and only poor timber. In nature we notice that there are many types of clay (nuclei of atom—types of element). One man offers to people to make from clay bricks (AB-Matter) and build from these bricks a fantastic array of desirable structures too complex to make from naturally occuring mounds of mud. The bricks enable by increased precision and strength things impossible before. A new level of human civilization begins.

The author calls upon scientists and the technical community to research and develop femtotechnology [10]. We can reach progress more quickly than in the further prospects of nanotechnology in this field, because we have fewer (only 3) initial components (proton, neutron, electron) and interaction between them is well-known (3 main forces: strong, weak, electrostatic). The different conventional atoms number about 100, most common molecules are tens of thousands and interactions between them are very complex (e.g. Van der Waals force).

What time horizon might we face in this quest? The physicist Richard Feynman offered his idea to design artificial matter from atoms and molecules at an American Physical Society meeting at Caltech on December 29, 1959. But only in the last 15 years we have initial progress in nanotechnology. On the other hand, progress is becoming swifter as more and better tools become common and as the technical community grows.

Now we are in the position of trying to progress from the ancient 'telega' haywagon of rural Russia (in analogy, conventional matter composites) to a 'luxury sport coupe' (advanced tailored nanomaterials). The author suggests we have little to lose and literal worlds to gain by simultaneously researching how to leap from 'telega' to 'hypersonic space plane'. (Femotech materials and technologies, enabling all the wonders outlined here) [1 – 10].

REFERENCES

(The reader may find some of these articles at the web storages: http://www.scribd.com , http://arxiv.org , http://vixra.org, http://aiaa.org , search "Bolonkin" and in the author's books: "*Non-Rocket Space Launch and Flight*", Elsevier, London, 2005, 488 pages; "*New Concepts, Ideas, Innovations in Aerospace, Technology and Human Science*", NOVA, 2006, 502 pages and "*Macro-Projects: Environment and Technology*", NOVA 2007, 536 pages; "New Technologies and Revolutionary Projects", Lulu, 2008, 324 pgs; Innovations and New Technologies. Lulu, 2013. 309 pgs. 8 Mb.).

[1] Bolonkin A.A., Femtotechnology. Nuclear AB-Matter with Fantastic Properties, *American Journal of Engineering and Applied Sciences*. 2 (2), 2009, p.501-514. Presented as paper AIAA-2009-4620 to 7th Annual International Energy Convention Conference, 2-5 August 2009, Denver, CO, USA. [On line]: http://www.scribd.com/doc/24045154. http://viXra.org/abs/1309.0201 .

[2] Bolonkin A.A., Femtotechnology: Design of the Strongest AB-Matter for Aerospace. Presented as paper AIAA-2009-4620 to 45 Joint Propulsion Conference, 2-5 August, 2009, Denver CO, USA. See also closed paper AIAA-2010-1556 in 48 Aerospace Meeting, New Horizons, 4 – 7 January, 2010, Orlando, FL, USA.
http://www.archive.org/details/FemtotechnologyDesignOfTheStrongestAb-matterForAerospace . Published in "Journal of Aerospace Engineering", Oct. 2010, Vol. 23, No. 4, pp.281-292.
http://www.archive.org/details/FemtotechnologyDesignOfTheStrongestAb-matterForAerospace
http://www.scribd.com/doc/57369206/Femtotechnology-Design-of-the-Strongest-AB-Matter-for-Aerospace

[3] Bolonkin A.A., AB-needles: Fantastic Properties and Application in Energy. International Journal of Energy Engineering (IJEE), December, 2013.

[4] Bolonkin A.A., Converting of Matter to Nuclear Energy by AB-Generator. *American Journal of Engineering and Applied Sciences*. 2 (4), 2009, p.683-693. [on line] http://www.scribd.com/doc/24048466. http://viXra.org/abs/1309.0200

[5] Bolonkin A.A., "Non-Rocket Space Launch and Flight", Elsevier, 2005, 488 pgs. ISBN-13: 978-0-08044-731- 5, ISBN-10: 0-080-44731-7 . http://www.archive.org/details/Non-rocketSpaceLaunchAndFlight, http://www.scribd.com/doc/24056182.

[6] Bolonkin A.A., "New Concepts, Ideas, Innovations in Aerospace, Technology and the Human Sciences", NOVA, 2006, 510 pgs. ISBN-13:978-1-60021-787-6.http://www.scribd.com/doc/24057071 , http://www.archive.org/details/NewConceptsIfeasAndInnovationsInAerospaceTechnologyAndHumanSciences, http://viXra.org/abs/1309.0193,

[7] Bolonkin A.A., Cathcart R.B., "Macro-Projects: Environments and Technologies", NOVA, 2007, 536 pgs. ISBN 978-1-60456-998-8. http://www.scribd.com/doc/24057930 . http://viXra.org/abs/1309.0192 , http://www.archive.org/details/Macro-projectsEnvironmentsAndTechnologies

[8] Bolonkin A.A., "New Technologies and Revolutionary Projects", Scribd, 2008, 324 pgs, http://www.scribd.com/doc/32744477, http://www.archive.org/details/NewTechnologiesAndRevolutionaryProjects.

[9] Bolonkin A.A., LIFE. SCIENCE. FUTURE (Biography notes, researches and innovations). Lambert, 2010, 208 pgs. 16 Mb. ISBN: 978-3-8473-0839-3 . http://www.scribd.com/doc/48229884, http://www.archive.org/details/Life.Science.Future.biographyNotesResearchesAndInnovations, http://viXra.org/abs/1309.0205 , http://www.lulu.com, search "Bolonkin".

[10] Bolonkin A.A., Femtotechnologies and Revolutionary Projects. Lambert, USA, 2011. 538 p. 16 Mb. ISBN: 978-3-8473-0839-0. http://www.scribd.com/doc/75519828/, ISBN 078-1-105-64111-4; http://viXra.org/abs/1309.0191, http://www.archive.org/details/FemtotechnologiesAndRevolutionaryProjects .

[11] http://www.physicum.narod,ru, Vol. 5., p.670.

[12] AIR Physics Desk Reference, Third Edition, p. 547.

13 November 2013

Possible structures from femtomatter

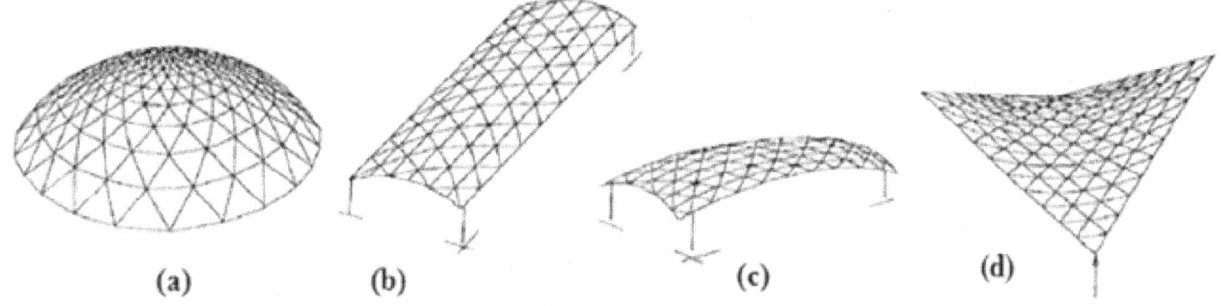

Size and scale of nucleus particles

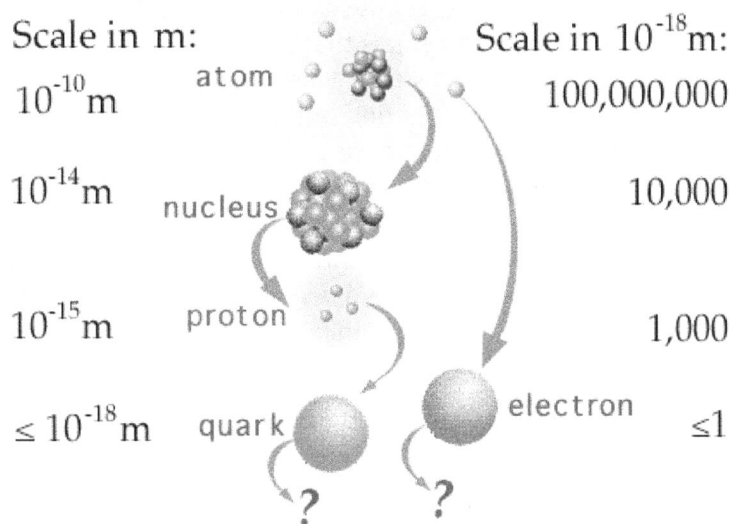

Chapter 5

Converting of Matter to Nuclear Energy

Abstract

Author offers a new nuclear generator which allows to convert any matter to nuclear energy in accordance with the Einstein equation $E=mc^2$. The method is based upon tapping the energy potential of a Micro Black Hole (MBH) and the Hawking radiation created by this MBH. As is well-known, the vacuum continuously produces virtual pairs of particles and antiparticles, in particular, the photons and anti-photons. The MBH event horizon allows separating them. Anti-photons can be moved to the MBH and be annihilated; decreasing the mass of the MBH, the resulting photons leave the MBH neighborhood as Hawking radiation. The offered nuclear generator (named by author as AB-Generator) utilizes the Hawking radiation and injects the matter into MBH and keeps MBH in a stable state with near-constant mass. The AB-Generator can not only produce gigantic energy outputs but should be hundreds of times cheaper than a conventional electric generation processes. The AB-Generator can be used in aerospace as a photon rocket or as a power source for numerous space vehicles. Many scientists expect the Large Hadron Collider at CERN will produce one MBH every second and the technology to capture them may be used for the AB-Generator.

Key words: Production of nuclear energy, Micro Black Hole, energy AB-Generator, photon rocket.
* Presented as Paper AIAA-2009-5342 in 45 Joint Propulsion Conferences, 2–5 August, 2009, Denver, CO, USA.

Introduction

Black hole. In general relativity, a black hole is a region of space in which the gravitational field is so powerful that nothing, including light, can escape its pull. The black hole has a one-way surface, called the event horizon, into which objects can fall, but out of which nothing can come out. It is called "black" because it absorbs all the light that hits it, reflecting nothing, just like a perfect blackbody in thermodynamics. Despite its invisible interior, a black hole can reveal its presence through interaction with other matter. A black hole can be inferred by tracking the movement of a group of stars that orbit a region in space which looks empty. Alternatively, one can see gas falling into a relatively small black hole, from a companion star. This gas spirals inward, heating up to very high temperature and emitting large amounts of radiation that can be detected from earthbound and earth-orbiting telescopes. Such observations have resulted in the general scientific consensus that, barring a breakdown in our understanding of nature, that black holes do exist in our universe. Although it is impossible to directly observe a black hole, its existence is inferred by its gravitational action on the surrounding environment, particularly with microquasars and active galactic nuclei, where material falling into a nearby black hole is significantly heated and emits a large amount of X-ray radiation. The only objects that agree with these observations and are consistent within the framework of general relativity are black holes.

A black hole has only three independent physical properties: mass, charge and angular momentum. In astronomy black holes are classed as:

- Supermassive - contain hundreds of thousands to billions of solar masses and are thought to exist in the center of most galaxies, including the Milky Way.

- Intermediate - contain thousands of solar masses.

- Micro (also *mini black holes*) - have masses much less than that of a star. At these sizes, quantum mechanics is expected to take effect. There is no known mechanism for them to form via normal processes of stellar evolution, but certain inflationary scenarios predict their production during the early stages of the evolution of the universe.

According to some theories of quantum gravity they may also be produced in the highly energetic reaction produced by cosmic rays hitting the atmosphere or even in particle accelerators such as the Large Hadron Collider. The theory of Hawking radiation predicts that such black holes will evaporate in bright flashes of gamma radiation. NASA's Fermi Gamma-ray Space Telescope satellite (formerly GLAST) launched in 2008 is searching for such flashes.

Fig 1. Artist's conception of a stellar mass black hole. Credit NASA.

The defining feature of a black hole is the appearance of an *event horizon*; a boundary in spacetime beyond which events cannot affect an outside observer. Since the event horizon is not a material surface but rather merely a mathematically defined demarcation boundary, nothing prevents matter or radiation from entering a black hole, only from exiting one.

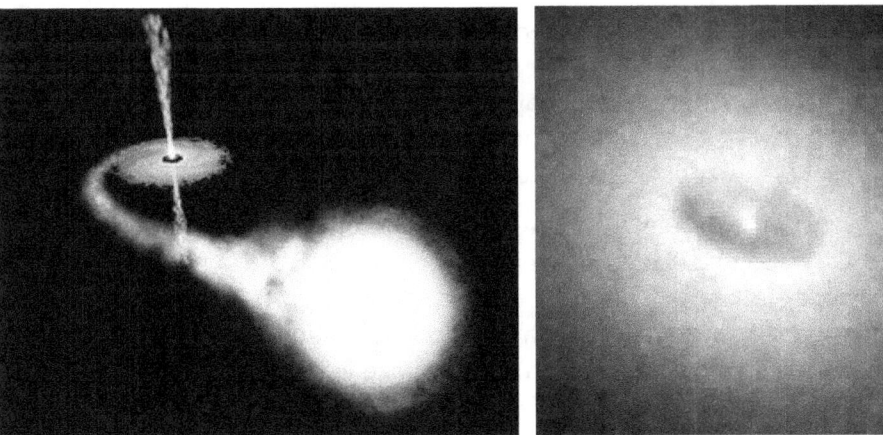

Fig.2 (left). Artist's impression of a binary system consisting of a black hole and a main sequence star. The black hole is drawing matter from the main sequence star via an accretion disk around it, and some of this matter forms a gas jet.
Fig.3 (right). Ring around a suspected black hole in galaxy NGC 4261. Date: Nov.1992. Courtesy of Space Telescope Science

For a non rotating (static) black hole, the *Schwarzschild radius* delimits a spherical event horizon. The Schwarzschild radius of an object is proportional to the mass. Rotating black holes have distorted, nonspherical event horizons. The description of black holes given by general relativity is known to be an approximation, and it is expected that quantum gravity effects become significant near the vicinity of the event horizon. This allows observations of matter in the vicinity of a black hole's event horizon to be used to indirectly study general relativity and proposed extensions to it.

Fig.4. Artist's rendering showing the space-time contours around a black hole. Credit NASA.

Though black holes themselves may not radiate energy, electromagnetic radiation and matter particles may be radiated from just outside the event horizon via *Hawking radiation*. At the center of a black hole lies the *singularity*, where matter is crushed to infinite density, the pull of gravity is infinitely strong, and spacetime has infinite curvature. This means that a black hole's mass becomes entirely compressed into a region with zero volume. This zero-volume, infinitely dense region at the center of a black hole is called a *gravitational singularity*. The singularity of a non-rotating black hole has zero length, width, and height; a rotating black hole's is smeared out to form a ring shape lying in the plane of rotation. The ring still has no thickness and hence no volume.

The *photon sphere* is a spherical boundary of zero thickness such that photons moving along tangents to the sphere will be trapped in a circular orbit. For non-rotating black holes, the photon sphere has a radius 1.5 times the Schwarzschild radius. The orbits are dynamically unstable, hence any small perturbation (such as a particle of infalling matter) will grow over time, either setting it on an outward trajectory escaping the black hole or on an inward spiral eventually crossing the event horizon.

Rotating black holes are surrounded by a region of spacetime in which it is impossible to stand still, called the *ergosphere*. Objects and radiation (including light) can stay in orbit within the ergosphere without falling to the center. Once a black hole has formed, it can continue to grow by absorbing additional matter. Any black hole will continually absorb interstellar dust from its direct surroundings and omnipresent cosmic background radiation. Much larger contributions can be obtained when a black hole merges with other stars or compact objects.

Hawking radiation. In 1974, Stephen Hawking showed that black holes are not entirely black but emit small amounts of thermal radiation.[1] He got this result by applying quantum field theory in a static black hole background. The result of his calculations is that a black hole should emit particles in a perfect black body spectrum. This effect has become known as Hawking radiation. Since Hawking's result many others have verified the effect through various methods. If his theory of black hole radiation is correct then black holes are expected to emit a thermal spectrum of radiation, and thereby lose mass, because according to the theory of relativity mass is just highly condensed energy ($E = mc^2$). Black holes will shrink and evaporate over time. The temperature of this spectrum (Hawking temperature) is proportional to the surface gravity of the black hole, which in turn is inversely proportional to the mass. Large black holes, therefore, emit less radiation than small black holes.

On the other hand if a black hole is very small, the radiation effects are expected to become very strong. Even a black hole that is heavy compared to a human would evaporate in an instant. A black hole the weight of a car (~10^{-24} m) would only take a nanosecond to evaporate, during which time it would briefly have a luminosity more than 200 times that of the sun. Lighter black holes are expected to evaporate even faster, for example a black hole of mass 1 TeV/c^2 would take less than 10^{-88} seconds to evaporate completely. Of course, for such a small black hole quantum gravitation effects are expected to play an important role and could even – although current developments in quantum gravity do not indicate so – hypothetically make such a small black hole stable.

Micro Black Holes. Gravitational collapse is not the only process that could create black holes. In principle, black holes could also be created in high energy collisions that create sufficient density. Since classically black holes can take any mass, one would expect micro black holes to be created in any such process no matter how low the energy. However, to date, no such events have ever been detected either directly or indirectly as a deficiency of the mass balance in particle accelerator experiments. This suggests that there must be a lower limit for the mass of black holes. Theoretically this boundary is expected to lie around the Planck mass (~10^{19} GeV/c^2, m_p = 2.1764·10^{-8} kg), where quantum effects are expected to make the theory of general relativity break down completely. This would put the creation of black holes firmly out of reach of any high energy process occurring on or near the Earth. Certain developments in quantum gravity however suggest that this bound could be much lower. Some braneworld scenarios for example put the Planck mass much lower, maybe even as low as 1 TeV. This would make it possible for micro black holes to be created in the high energy collisions occurring when cosmic rays hit the Earth's atmosphere, or possibly in the new Large Hadron Collider at CERN. These theories are however very speculative, and the creation of black holes in these processes is deemed unlikely by many specialists.

Smallest possible black hole. To make a black hole one must concentrate mass or energy sufficiently that the escape velocity from the region in which it is concentrated exceeds the speed of light. This condition gives the Schwarzschild radius, $r_o = 2GM/c^2$, where G is Newton's constant and c is the speed of light, as the size of a black hole of mass M. On the other hand, the Compton wavelength, $\lambda = h/Mc$, where h is Planck's constant, represents a limit on the minimum size of the region in which a mass M at rest can be localized. For sufficiently small M, the Compton wavelength exceeds the Schwarzschild radius, and no black hole description exists. This smallest mass for a black hole is thus approximately the Planck mass, which is about 2×10^{-8} kg or 1.2×10^{19} GeV/c^2.

Any primordial black holes of sufficiently low mass will Hawking evaporate to near the Planck mass within the lifetime of the universe. In this process, these small black holes radiate away matter. A rough picture of this is that pairs of virtual particles emerge from the vacuum near the event horizon, with one member of a pair being captured, and the other escaping the vicinity of the black hole. The net result is the black hole loses mass (due to conservation of energy). According to the formulae of black hole thermodynamics, the more the black hole loses mass the hotter it becomes, and the faster it evaporates, until it approaches the Planck mass. At this stage a black hole would have a Hawking temperature of $T_P/8\pi$ (5.6×10^{32} K), which means an emitted Hawking particle would have an energy comparable to the mass of the black hole. Thus a thermodynamic description breaks down. Such a mini-black hole would also have an entropy of only 4π nats, approximately the minimum possible value.

At this point then, the object can no longer be described as a classical black hole, and Hawking's calculations also break down. Conjectures for the final fate of the black hole include total evaporation and production of a Planck mass-sized *black hole remnant*. If intuitions about quantum black holes are correct, then close to the Planck mass the number of possible quantum states of the black hole is expected to become so few and so quantised that its interactions are likely to be quenched out. It is possible that such Planck-mass black holes, no longer able either to absorb energy gravitationally like a classical black hole because of the quantised gaps between their allowed energy levels, nor to emit Hawking particles for the same reason, may in effect be stable objects. They would in effect be WIMPs, weakly interacting massive particles; this could explain dark matter.

Creation of micro black holes. Production of a black hole requires concentration of mass or energy within the corresponding Schwarzschild radius. In familiar three-dimensional gravity, the minimum such energy is 10^{19} GeV, which would have to be condensed into a region of approximate size 10^{-33} cm. This is far beyond the limits of any current technology; the Large hadron collider (LHC) has a design energy of 14 TeV. This is also beyond the range of known collisions of cosmic rays with Earth's atmosphere, which reach center of mass energies in the range of hundreds of TeV. It is estimated that to collide two particles to within a distance of a Planck length with currently achievable magnetic field strengths would require a ring accelerator about 1000 light years in diameter to keep the particles on track.

Some extensions of present physics posit the existence of extra dimensions of space. In higher-dimensional spacetime, the strength of gravity increases more rapidly with decreasing distance than in three dimensions. With certain special configurations of the extra dimensions, this effect can lower the Planck scale to the TeV range. Examples of such extensions include large extra dimensions, special cases of the Randall-Sundrum model, and String theory configurations. In such scenarios, black hole production could possibly be an important and observable effect at the LHC.

Virtual particles. In physics, a virtual particle is a particle that exists for a limited time and space, introducing uncertainty in their energy and momentum due to the Heisenberg Uncertainty Principle. Vacuum energy can also be thought of in terms of virtual particles (also known as vacuum fluctuations) which are created and destroyed out of the vacuum. These particles are always created out of the vacuum in particle-antiparticle pairs, which shortly annihilate each other and disappear. However, these particles and antiparticles may interact with others before disappearing.

The net energy of the Universe remains zero so long as the particle pairs annihilate each other within Planck time. Virtual particles are also excitations of the underlying fields, but are detectable only as forces. The creation of these virtual particles near the event horizon of a black hole has been hypothesized by physicist Stephen Hawking to be a mechanism for the eventual "evaporation" of black holes. Since these particles do not have a permanent existence, they are called *virtual particles* or vacuum fluctuations of vacuum energy. An important example of the "presence" of virtual particles in a vacuum is the Casimir effect. Here, the explanation of the effect requires that the total energy of all of the virtual particles in a vacuum can be added together. Thus, although the virtual particles themselves are not directly observable in the laboratory, they do leave an observable effect: their zero-point energy results in forces acting on suitably arranged metal plates or dielectrics. Thus, virtual particles are often popularly described as coming in pairs, a particle and antiparticle, which can be of any kind.

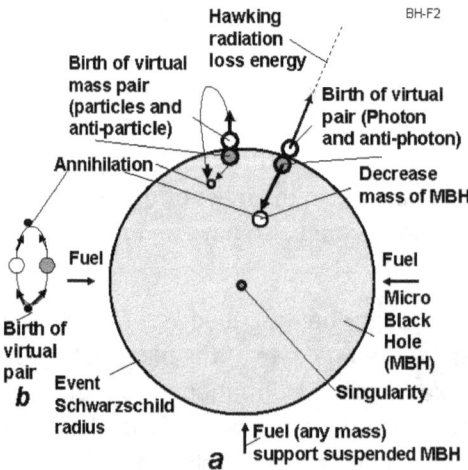

Fig.5. Hawking radiation. *a*. Virtual particles at even horizon.
b. Virtual particles out even horizon (in conventional space).

The evaporation of a black hole is a process dominated by photons, which are their own antiparticles and are uncharged. The uncertainty principle in the form $\Delta E \Delta t \geq h$ implies that in the vacuum one or more particles with energy ΔE above the vacuum may be created for a short time Δt. These *virtual particles* are included in the definition of the vacuum.

Vacuum energy is an underlying background energy that exists in space even when devoid of matter (known as free space). The vacuum energy is deduced from the concept of virtual particles, which are themselves derived from the energy-time uncertainty principle. Its effects can be observed in various phenomena (such as spontaneous emission, the Casimir effect, the van der Waals bonds, or the Lamb shift), and it is thought to have consequences for the behavior of the Universe on cosmological scales.

AB-Generator of Nuclear Energy and some Innovations

Simplified explanation of MBH radiation and work of AB-Generator (Fig.5). As known, the vacuum continuously produces, virtual pairs of particles and antiparticles, in particular, photons and anti-photons. In conventional space they exist only for a very short time, then annihilate and return back to nothingness. The MBH event horizon, having very strong super-gravity, allows separation of the particles and anti particles, in particular, photons and anti-photons. Part of the anti-photons move into the MBH and annihilate with photons decreasing the mass of the MBH and return back a borrow energy to vacuum. The free photons leave from the MBH neighborhood as Hawking radiation. That way the MBH converts any conventional matter to Hawking radiation which may be converted to heat or electric energy by the AB- Generator. This AB- Generator utilizes the produced Hawking radiation and injects the matter into the MBH while maintaining the MBH in stable suspended state.

Note: The photon does NOT have rest mass. Therefore a photon can leave the MBH's neighborhood (if it is located beyond the event horizon). All other particles having a rest mass and speed less than light speed *cannot* leave the Black Hole. They cannot achieve light speed because their mass at light speed equals infinity and requests infinite energy for its' escape—an impossibility.

Description of AB- Generator. The offered nuclear energy AB- Generator is shown in fig. 6. That includes the Micro Black Hole (MBH) 1 suspended within a spherical radiation reflector and heater 5. The MBH is supported (and controlled) at the center of sphere by a fuel (plasma, proton, electron, matter) gun 7. This AB- Generator also contains the 9 – heat engine (for example, gas, vapor turbine), 10 – electric generator, 11 – coolant (heat transfer agent), an outer electric line 12, internal electric generator (5 as antenna) with customer 14.

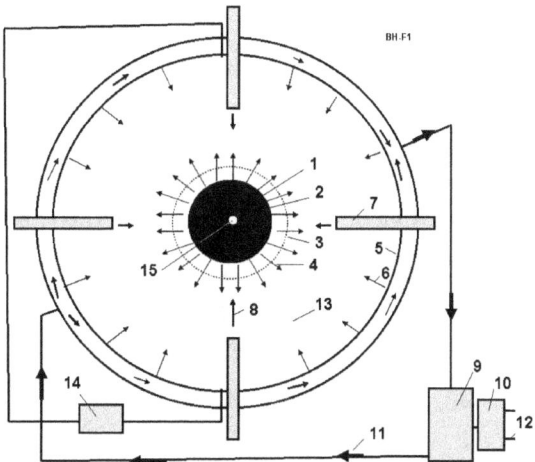

Fig.6. Offered **nuclear-vacuum energy AB- Generator**. *Notations*: 1- Micro Black Hole (MBH), 2 - event horizon (Schwarzschild radius), 3 - photon sphere, 4 – black hole radiation, 5 – radiation reflector, antenna and heater (cover sphere), 6 – back (reflected) radiation from radiation reflector 5, 7 – fuel (plasma, protons, electrons, ions, matter) gun

(focusing accelerator), 8 – matter injected to MBH (fuel for Micro Black hole), 9 – heat engine (for example, gas, vapor turbine), 10 – electric generator connected to heat engine 9, 11 – coolant (heat transfer agent to the heat machine 9), 12 – electric line, 13 – internal vacuum, 14 – customer of electricity from antenna 5, 15 – singularity.

Work. The generator works the following way. MBH, by selective directional input of matter, is levitated in captivity and produces radiation energy 4. That radiation heats the spherical reflector-heater 5. The coolant (heat transfer agent) 11 delivers the heat to a heat machine 9 (for example, gas, vapor turbine). The heat machine rotates an electric generator 10 that produces the electricity to the outer electric line 12. Part of MBH radiation may accept by sphere 5 (as antenna) in form of electricity.

The control fuel guns inject the matter into MBH and do not allow bursting of the MBH. This action also supports the MBH in isolation, suspended from dangerous contact with conventional matter. They also control the MBH size and the energy output.

Any matter may be used as the fuel, for example, accelerated plasma, ions, protons, electrons, micro particles, etc. The MBH may be charged and rotated. In this case the MBH may has an additional suspension by control charges located at the ends of fuel guns or (in case of the rotating charged MBH) may have an additional suspension by the control electric magnets located on the ends of fuel guns or at points along the reflector-heater sphere.

Innovations, features, advantages and same research results

Some problems and solutions offered by the author include the following:

1) A practical (the MBH being obtained and levitated, details of which are beyond the scope of this paper) method and installation for converting any conventional matter to energy in accordance with Einstein's equation $E = mc^2$.
2) MBHs may produce gigantic energy and this energy is in the form of dangerous gamma radiation. The author shows how this dangerous gamma radiation Doppler shifts when it moves against the MBH gravity and converts to safely tapped short radio waves.
3) The MBH of marginal mass has a tendency to explode (through quantum evaporation, very quickly radiating its mass in energy). The AB- Generator automatically injects metered amounts of matter into the MBH and keeps the MGH in a stable state or grows the MBH to a needed size, or decreases that size, or temporarily turns off the AB- Generator (decreases the MBH to a Planck Black Hole).
4) Author shows the radiation flux exposure of AB- Generator (as result of MBH exposure) is not dangerous because the generator cover sphere has a vacuum, and the MBH gravity gradient decreases the radiation energy.
5) The MBH may be supported in a levitated (non-contact) state by generator fuel injectors.

Theory of AB- Generator

Below there are main equations for computation the conventional black hole (BH) and AB-Generator

General theory of Black Hole.

1. Power produced by BH is

$$P = \frac{\hbar c^6}{15360\pi G^2} \frac{1}{M} \approx 3.56 \cdot 10^{32} \frac{1}{M}, \quad W, \qquad (1)$$

where $\hbar = h/2\pi = 1.0546 \cdot 10^{-34}$ J/s is reduced Planck constant, $c = 3 \cdot 10^8$ m/s - light speed, $G = 6.6743 \cdot 10^{-11}$ m³/kg.s² is gravitation constant, M – mass of BH, kg.

2. Temperature of black body corresponding to this radiation is

$$T = \frac{hc^3}{8\pi G k_b} \frac{1}{M} \approx 1.23 \cdot 10^{23} \frac{1}{M}, \quad K, \tag{2}$$

where $k_b = 1.38 \cdot 10^{-23}$ J/k is Boltzmann constant.

3. Energy E_p [J] and frequency ν_o of photon at event horizon are

$$E_p = \frac{hc^3}{16\pi G} \frac{1}{M}, \quad \nu_0 = \frac{E_p}{h} = \frac{c^3}{16\pi G} \frac{1}{M} = 8.037 \cdot 10^{33} \frac{1}{M}, \quad \lambda_0 = \frac{c}{\nu_0} = 3.73 \cdot 10^{-26} M. \tag{3}$$

where $c = 3 \cdot 10^8$ m/s is light speed, λ_o is wavelength of photon at even radius, m. h is Planck constant.

4. Radius of BH event horizon (Schwarzschild radius) is

$$r_0 = \frac{2G}{c^2} M = 1.48 \cdot 10^{-27} M, \quad m, \tag{4}$$

5. Relative density (ratio of mass M to volume V of BH) is

$$\rho = \frac{M}{V} = \frac{3c^2}{32\pi G^3} \frac{1}{M^2} \approx 7.33 \cdot 10^{79} \frac{1}{M^2}, \quad kg/m^3. \tag{5}$$

6. Maximal charge of BH is

$$Q_{max} = 5 \cdot 10^9 eM \approx 8 \cdot 10^{-10} M, \quad C, \tag{6}$$

where $e = -1.6 \cdot 10^{-19}$ is charge of electron, C.

7. Life time of BH is

$$\tau = \frac{5120\pi G^2}{hc^4} M^3 = 2.527 \cdot 10^{-8} M^3, \quad s. \tag{7}$$

8. Gravitation around BH (r is distance from center) and on event horizon

$$g = \frac{GM}{r^2}, \quad g_0 = \frac{c^4}{4G} \frac{1}{M} = 3 \cdot 10^{42} \frac{1}{M}, \quad m\,s^{-2}. \tag{8}$$

Developed Theory of AB- Generator

Below are research and the theory developed by author for estimation and computation of facets of the AB-Generator.

9. Loss of energy of Hawking photon in BH gravitational field. It is known that a red shift allows estimating the frequency of photon in central gravitational field when it moves TO the gravity center. In this case the photon increases its frequency because photon is accelerated the gravitational field (wavelength decreases). But in our case the photon moves FROM the gravitational center, the gravitational field brakes it and the photon loses its energy. That means its frequency decreases and the wavelength increases. Our photon gets double energy because the black hole annihilates two photons (photon and anti-photon). That way the equation for photon frequency at distance $r > r_o$ from center we can write in form

$$\frac{v}{v_0} \approx 1 + \frac{2\Delta\varphi}{c^2}, \qquad (9)$$

Where Δϕ = ϕ − ϕ₀ is difference of the gravity potential. The gravity potential is

$$\Delta\varphi = \varphi - \varphi_0, \quad \varphi = \frac{GM}{r}, \quad \varphi_0 = \frac{GM}{r_0}, \quad r_0 = \frac{2GM}{c^2}. \qquad (10)$$

Let us substitute (10) in (9), we get

$$\frac{v}{v_0} \approx 1 + \frac{r_0}{r} - \frac{r_0}{r_0}, \quad \text{or} \quad \frac{v}{v_0} = \frac{\lambda_0}{\lambda} \approx \frac{r_0}{r}. \qquad (11)$$

It is known, the energy and mass of photon is

$$E_f = h\gamma, \quad E_f = m_f c^2, \quad m_f = E_f / c^2, \qquad (12)$$

The energy of photon linear depends from its frequency. Reminder: The photon does not have a rest mass.

The relative loss of the photon radiation energy ξ at distance r from BH and the power P_r of Hawking radiation at radius r from the BH center is

$$\xi = \frac{r_0}{r}, \quad v = \xi v_0, \quad P_r = \xi P. \qquad (13)$$

The r_0 is very small and ξ is also very small and $v \ll v_0$.

The result of an energy loss by Hawking photon in the BH gravitational field is very important for AB-Generator. The energy of Hawking radiation is very big; we very need to decrease it in many orders. The initial Hawking photon is gamma radiation that is dangerous for people and matter. In r distance the gamma radiation may be converted in the conventional light or radio radiation, which are not dangerous and may be reflected, focused or a straightforward way converted into electricity by antenna.

10. Reflection Hawking radiation back to MBH. For further decreasing the MBH produced energy the part of this energy may be reflected to back in MBH. A conventional mirror may reflect up 0.9 ÷ 0.99 of radiation (ξ_r = 0.01 ÷ 0.1, ξ_r is a loss of energy in reflecting), the multi layers mirror can reflect up 0.9999 of the monochromatic light radiation ($\xi_r = 10^{-3} \div 10^{-5}$), and AB-mirror from cubic corner cells offered by author in [2], p. 226, fig.12.1g , p. 376 allows to reflect non-monochromatic light radiation with efficiency up $\xi_r = 10^{-13}$ strong back to source. In the last case, the loss of reflected energy is ([2] p.377)

$$\xi_r = 0.00023 al, \quad l = m\lambda, \quad m \geq 1, \qquad (14)$$

where l is size of cube corner cell, m; m is number of radiation waves in one sell; λ is wavelength, m; a is characteristic of sell material (see [2], fig.A3.3). Minimal value $a = 10^{-2}$ for glass and $a = 10^{-4}$ for KCl crystal.

The reflection of radiation to back in MBH is may be important for MBH stabilization, MBH storage and MBH 'switch off'.

11. Useful energy of AB- Generator. The useful energy P_u [J] is taken from AB- Generator is

$$P_u = \xi \xi_r P. \qquad (15)$$

12. Fuel consumption is

$$\overset{8}{M} = P_u / c^2, \quad \text{kg.} \qquad (16)$$

The fuel consumption is very small. AB- Generator is the single method in the World now known which allows full converting reasonably practical conversion of (any!) matter into energy according the Einsteinian equation $E = mc^2$.

13. Specific pressure on AB- Generator cover sphere p [N/m²] **and on the surface of MBH** p_o **is**

$$p = \frac{kP_r}{Sc} = \frac{kP_r}{4\pi r^2 c} = 2.65 \cdot 10^{-10} \frac{kP_r}{r^2}, \quad p_0 = \frac{P}{S_0 c} = \frac{hc^8}{15360 \cdot 16\pi^2 G^4} \frac{1}{M^4} = 1.44 \cdot 10^{28} \frac{1}{M^4}, \quad (17)$$

where $k = 1$ if the cover sphere absorbs the radiation and $k \approx 2$ if the cover sphere high reflects the radiation, S is the internal area of cover sphere, m²; S_0 is surface of event horizon sphere, m²; p_o is specific pressure of Hawking radiation on the event horizon surface. Note, the pressure p on cover sphere is small (see Project), but pressure p_o on event horizon surface is very high.

14. Mass particles produced on event surface. On event horizon surface may be also produced the mass particles with speed $V < c$. Let us take the best case (for leaving the BH) when their speed is radially vertical. They cannot leave the BH because their speed V is less than light speed c. The maximal radius of lifting r_m [m] is

$$dV = -g dt, \quad dV = -\frac{g}{V} dr = -\frac{GM}{V} \frac{dr}{r^2}, \quad r_m = \frac{2GM}{c^2 - V_0^2} = \frac{r_0}{1 - (V/c)^2}, \quad (18)$$

where g is gravitational acceleration of BH, m/s²; t is time, sec.; r_o is BH radius, m; V_0 is particle speed on event surface, m/s². If the r_m is less than radius of the cover sphere, the mass particles return to BH and do not influence the heat flow from BH to cover sphere. That is in the majority of cases.

15. Explosion of MBH. The MBH explosion produces the radiation energy

$$E_e = Mc^2. \quad (19)$$

MBH has a small mass. The explosion of MBH having $M = 10^{-5}$ kg produces 9×10^{11} J. That is energy of about 10 tons of good conventional explosive (10^7 J/kg). But there is a vacuum into the cover sphere and this energy is presented in radiation form. But in reality only very small part of explosion energy reaches the cover sphere, because the very strong MBH gravitation field brakes the photons and any mass particles. Find the energy which reaches the cover sphere via:

$$dE = \xi c^2 dM, \quad \xi = \frac{r_0}{r}, \quad r_0 = \frac{2G}{c^2} M, \quad dE = \frac{2G}{r} M dM, \quad E = \frac{G}{r} M^2 = 6{,}674 \cdot 10^{-11} \frac{M^2}{r}. \quad (20)$$

The specific exposure radiation pressure of MBH pressure p_e [N/m²] on the cover sphere of radius $r < r_0$ may be computed by the way:

$$p_e = \frac{E}{V} = \frac{3G}{4\pi} \frac{M^2}{r^3} = 1.6 \cdot 10^{-11} \frac{M^2}{r^3}, \quad r > r_0, \quad (21)$$

where $V = 3/4 \pi r^3$ is volume of the cover sphere.

That way the exposure radiation pressure on sphere has very small value and presses very short time. Conventional gas balloon keeps pressure up 10^7 N/m² (100 atm). However, the heat impact may be high and AB- Generator design may have the reflectivity cover and automatically open windows for radiation.

Your attention is requested toward the next important result following from equations (20)-(21). Many astronomers try to find (detect) the MBH by a MBH exposure radiation. But this radiation is small, may be detected but for a short distance, does not have a specific frequency and has a variably long wavelength. This may be why during more than 30 years nobody has successfully observed MBH events in Earth environment though the theoretical estimation predicts about 100 of MBH events annually. Observers take note!

16. Supporting the MBH in suspended (levitated) state. The fuel injector can support the MBH in suspended state (no contact the MBH with any material surface).

The maximal suspended force equals

$$F = qV_f, \quad q = \frac{P_u}{c^2}, \quad F = \frac{P_u V_f}{c^2}, \qquad (22)$$

where q is fuel consumption, kg; V_f is a fuel speed, m/s. The fuel (plasma) speed $0.01c$ is conventionally enough for supporting the MBH in suspended state.

17. AB- Generator as electric generator. When the Hawking radiation reaches the cover as radio microwaves they may be straightforwardly converted to electricity because they create a different voltage between different isolated parts of the cover sphere as in an antenna. Maximal voltage which can produces the radiation wave is

$$w = \frac{\varepsilon \varepsilon_0 E^2}{2} + \frac{\mu \mu_0 H^2}{2}, \quad w = \frac{P_r}{c}, \qquad (23)$$

where w is density of radiation energy, J/m³; E is electric intensity, V/m; H is magnetic intensity, T; $\varepsilon_0 = 8.85 \times 10^{-12}$ F/m is the coefficient of the electric permeability; $\mu_0 = 4\pi \times 10^{-7}$ N/A² is the coefficient of the magnetic permeability; $\varepsilon = \mu = 1$ for vacuum.

Let us take moment when $H = 0$, then

$$E = \sqrt{\frac{2w}{\varepsilon_0}} = \sqrt{\frac{2P_r}{\varepsilon_0 c}} = 2.73\sqrt{P_r} \quad U \approx bDE, \quad b = \frac{D}{\lambda} \leq 1,$$

$$P_e \approx bP_r, \quad \lambda = \lambda_0 \frac{r}{r_0} = 16r, \quad b = \frac{2r}{16r} = \frac{1}{8}, \qquad (24)$$

where E is electric intensity, V/m; U is voltage of AB-generator, V; b is relative size of antenna, D is diameter of the cover sphere if the cover sphere is used as a full antenna, m; P_e is power of the electric station, W.

As you see about 1/8 of total energy produced by AB- Generator we can receive in the form of electricity and 7/8ths reflects back to MBH; we may tap heat energy which convert to any form of energy by conventional (heat engine) methods. If we reflect the most part of the heat energy back into the MBH, we can have only electricity and do not have heat flux.

If we will use the super strong and super high temperature material AB-material offered in [3] the conversion coefficient of heat machine may be very high.

18. Critical mass of MBH located in matter environment. Many people are afraid the MBH experiments because BH can absorb the Earth. Let us find the critical mass of MBH which can begin uncontrollably to grow into the Earth environment. That will happen when BH begins to have more mass than mass of Hawking radiation. Below is the equation for the critical mass of initial BH. The educated reader will understand the equations below without detailed explanations.

$$dV = gdt, \quad g = \frac{GM}{r^2}, \quad dt = \frac{dr}{V}, \quad VdV = gdr, \quad \int_V^c VdV = \int_r^{r_0} \frac{GM}{r^2} dr, \quad r_0 = \frac{2G}{c^2}M, \quad V^2 = c^2\frac{r_0}{r},$$

$$V = c\sqrt{\frac{r_0}{r}}, \quad dt = \frac{\sqrt{r}dr}{c\sqrt{r_0}}, \quad \int_t^0 dt = \frac{1}{c\sqrt{r_0}} \int_r^{r_0} \sqrt{r}dr, \quad t = \frac{2}{3c\sqrt{r_0}}(r^{3/2} - r_0^{3/2}) \approx \frac{2r^{3/2}}{3cr_0^{1/2}}, \quad r = \left(\frac{3c\sqrt{r_0}}{2}t\right)^{3/2}, \qquad (25)$$

$$r = 1.65 G^{1/2} M^{1/3} t^{2/3}, \quad \dot{M} = \frac{P}{c^2} = \frac{hc^4}{15360\pi G^2} \frac{1}{M^2} = 4 \cdot 10^{15} \frac{1}{M^2}, \quad \text{for} \quad t = 1 \text{ s},$$

$$\dot{M}_e = \frac{4}{3}\pi r^3 \gamma = 6\pi \gamma G^{3/2} M \approx 10^{-4} \gamma M, \quad M = M_c e^{6\pi \gamma G^{3/2} t} \approx M_c e^{10^{-4} \gamma t}, \quad t = \frac{1}{6\pi \gamma G^{3/2}} \ln \frac{M}{M_c} \approx \frac{10^4}{\gamma} \ln \frac{M}{M_c},$$

where V is speed of environment matter absorbed by MBH, m/s; g is gravity acceleration of MBH, m/s; r is distance environment matter to MBH center, m; *t* is time, sec; $\dot{M}$ is mass loss by MBH, kg; $\dot{M}_e$ is mass taken from Earth environment by MBH, kg; *γ* is density of Earth environment, kg/m³; M_c is critical mass of MBH when one begin uncontrollable grows, kg; *t* is time, sec.

Let us to equate the mass $\dot{M}$ radiated by MBH to mass $\dot{M}_e$ absorbed by MBH from Earth environment, we obtain the critical mass M_c of MBH for any environment:

$$M_c^3 = \frac{hc^4}{92160\pi^2 G^3}\frac{1}{\gamma} = 3.17 \cdot 10^{24}\frac{1}{\gamma}, \quad \text{or} \quad \gamma = 3.17 \cdot 10^{24}\frac{1}{M_c^3}, \tag{26}$$

If MBH having mass $M = 10^7$ kg (10 thousands tons) is put in water (*γ* = 1000 kg/m³), this MBH can begin uncontrollable runaway growth and in short time (~74 sec) can consume the Earth into a black hole having diameter ~ 9 mm. If this MBH is located in the sea level atmosphere (*γ* = 1.29 kg/m³), the initial MBH must has critical mass $M = 10^8$ kg (100 thousand tons). The critical radius of MBH is very small. In the first case ($M = 10^7$ kg) $r_o = 1.48 \times 10^{-20}$ m, in the second case ($M = 10^8$ kg) $r_o = 1.48 \times 10^{-19}$ m. Our MBH into AB-Generator is not dangerous for Earth because it is located in vacuum and has mass thousands to millions times less than the critical mass.

However, in a moment of extreme speculation, if far future artificial intelligence (or super-small reasoning) beings will be created from nuclear matter [3] they can convert the Earth into a black hole to attempt to access quick travel to other stars (Solar systems), past and future Universes and even possibly past and future times.

19. General note. We got our equations in assumption $\lambda/\lambda_o = r/r_o$. If $\lambda/\lambda_o = (r/r_o)^{0.5}$ or other relation, the all above equations may be easy modified.

AB-Generator as Photon Rocket

The offered AB- Generator may be used as the most efficient photon propulsion system (photon rocket). The photon rocket is the dream of all astronauts and space engineers, a unique vehicle) which would make practical interstellar travel. But a functioning photon rocket would require gigantic energy. The AB- Generator can convert any matter in energy (radiation) and gives the maximum theoretical efficiency.

The some possible photon propulsion system used the AB –Generator is shown in Fig.7. In simplest version (***a***) the cover of AB generator has window 3, the radiation goes out through window and produces the thrust. More complex version (***c***) has the parabolic reflector, which sends all radiation in one direction and increases the efficiency. If an insert in the AB- Generator covers the lens 6 which will focuses the radiation in a given direction, at the given point the temperature will be a billions degree (see Equation (2)) and AB- Generator may be used as a photon weapon.

The maximal thrust *T* of the photon engine having AB- Generator may be computed (estimated) by equation:

$$T = \dot{M}c, \quad \text{N}, \tag{26}$$

For example, the AB-generator, which spends only 1 gram of matter per second, will produce a thrust 3×10⁵ N or 30 tons.

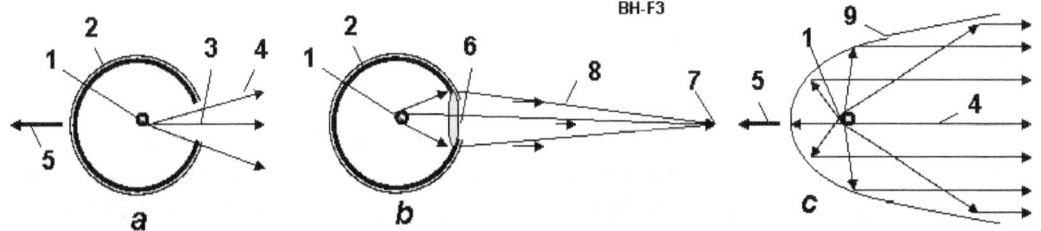

Fig.7. AB- Generator as Photon Rocket and Radiation (Photon) Weapon. (***a***) AB- Generator as a Simplest Photon Rocket;

(*b*) AB- Generator as focused Radiation (photon, light or laser) weapon; (*c*) Photon Rocket with Micro-Black Hole of AB-Generator. *Notations*: 1 – control MBH; 2 – spherical cover of AB-Generator; 3 – window in spherical cover; 4 – radiation of BH; 5 – thrust; 6 – lens in window of cover; 7 – aim; 8 - focused radiation; 9 – parabolic reflector.

AB-Generator Energy Production

To estimate the energy production of an AB-Generator which is only by way of example of a computation and possible parameters. Let us take the MBH mass $M = 10^{-5}$ kg and radius of the cover sphere $r = 5$ m. No reflection. Using the equations (1)-(24) we receive:

$$P = 3.56 \cdot 10^{32} / M^2 = 3,56 \cdot 10^{42} \quad \text{W},$$

$$r_0 = 1.48 \cdot 10^{-27} M = 1.48 \cdot 10^{-32} \quad \text{m},$$

$$\xi = r_0 / r = 2.96 \cdot 10^{-33},$$

$$P_r = \xi P = 1.05 \cdot 10^{10}, \quad P_u = \xi \xi_r P = P_r, \quad \text{W}, \quad \xi_r = 1.$$

$$\lambda_0 = 3,73 \cdot 10^{-26} M = 3.73 \cdot 10^{-31} \quad \text{m}. \tag{27}$$

$$\lambda = 16 \cdot r = 80 \quad \text{m}.$$

$$p = \frac{P_r}{4\pi c r^2} = 0.111 \quad \frac{N}{m^2}, \quad c = 3 \cdot 10^8 \quad \text{m/s},$$

$$\dot{M} = P_u / c^2 = 1.17 \cdot 10^{-7} \quad \text{kg/s},$$

$$p_e = 1.6 \cdot 10^{-11} \frac{M^2}{r^3} = 1.28 \cdot 10^{-23} \quad \text{N/m}^2$$

Remaing main notations in equations (27): $P_r = P_u = 1.05 \times 10^{10}$ W is the useful energy (1/8 of this energy may be taken as electric energy by cover antenna, 7/8 is taken as heat); $\lambda = 80$ m is wavelength of radiation at cover sphere (that is not dangerous for people); $\dot{M} = 1.17 \times 10^{-7}$ kg/s is fuel consumption; $r_0 = 1.48 \times 10^{-32}$ m is radius of MBH; $p_e = 1.28 \times 10^{-23}$ N/m² is explosion pressure of MBH.

Note that pressure of the explosion pressure is very small, less than a billion times of radiation pressure on the cover surface $p = 0.111$ N/m² which is not surprising because BH takes back the energy with that spent for acceleration the matter in eating the matter. As such, there is no danger of explosion of MBH.

Heat transfer and internal electric power are

$$q = \frac{P_u}{S} = \frac{P_u}{4\pi r^2} = 3.34 \cdot 10^7 \quad \frac{W}{m^2},$$

For $\delta = 2 \cdot 10^{-3}$ m, $\lambda_h = 100$, $\Delta T \approx q\delta / \lambda_h = 668°$K, \hfill (28)

$$E = 2.73\sqrt{P_r} = 2.8 \cdot 10^5 \quad \text{V/m}, \quad U = E \cdot 2r = 2.8 \cdot 10^6 \quad \text{V}, \quad P_e = P_r / 8 = 1.31 \cdot 10^9 \quad \text{W},$$

where q is specific heat transfer through the cover sphere, S is internal surface of the cover sphere, m²; δ is thickness of the cover sphere wall, m; λ_h is heat transfer coefficient for steel; ΔT is difference temperature between internal and external walls of the cover sphere; E is electric intensity from radiation on cover sphere surface, V/m; U is maximal electric voltage, V; P_e is electric power, W.

The power heat and electric output of a AB- Generator as similar to a very large complex of present day Earth's electric power stations ($P_r = 10^{10}$ W, ten billion of watts). The AB- Generator is a hundred times cheaper than a conventional electric station, especially since, heat energy can be reflected back to the MBH avoiding all the problems of conventional power conversion equipment (using only electricity from spherical cover as antenna). We hope the Large Hadron Collider at CERN can get the initial MBH needed for AB-Generator. The other way to obtain one is to find the Planck MBH (remaining from the time of the Big Bang and former MBH) and grow them to target MBH size.

Results

1. Author has offered the method and installation for converting any conventional matter to energy according the Einstein's equation $E = mc^2$, where m is mass of matter, kg; $c = 3 \cdot 10^8$ is light speed, m/s.
2. The Micro Black Hole (MBH) is offered for this conversion.
3. Also is offered the control fuel guns and radiation reflector for explosion prevention of MBH.
4. Also is offered the control fuel guns and radiation reflector for the MBH control.
5. Also is offered the control fuel guns and radiation reflector for non-contact suspension (levitation) of the MBH.
6. For non contact levitation of MBH the author also offers:
   a) Controlled charging of MBH and of ends of the fuel guns.
   b) Control charging of rotating MBH and control of electric magnets located on the ends of the fuel guns or out of the reflector-heater sphere.
7. The author researches show the very important fact: A strong gamma radiation produced by Hawking radiation loses energy after passing through the very strong gravitational MBH field. The MBH radiation can reach the reflector-heater as the light or short-wave radio radiation. That is very important for safety of the operating crew of the AB- Generator.
8. The author researches show: The matter particles produced by the MBH cannot escape from MBH and can not influence the Hawking radiation.
9. The author researches show another very important fact: The MBH explosion (hundreds and thousands of TNT tons) in radiation form produces a small pressure on the reflector-heater (cover sphere) and does not destroys the AB- Generator (in a correct design of AB-generator!). That is very important for safety of the operating crew of the AB-generator.
10. The author researches show another very important fact: the MBH cannot capture by oneself the surrounding matter and cannot automatically grow to consume the planet.
11. As the initial MBH can be used the Planck's (quantum) MBH which *may* be everywhere. The offered fuel gun may to grow them (or decrease them) to needed size or the initial MBH may be used the MBH produce Large Hadron Collider (LHC) at CERN. Some scientists assume LHC will produce one MBH every second (86,400 MBH in day). The cosmic radiation also produces about 100 MBH every year.
12. The spherical dome of MBH may convert part of the radiation energy to electricity.
13. A correct design of MBH generator does not produce the radioactive waste of environment.
14. The attempts of many astronomers find (detect) the MBH by a MBH exposure radiation will not be successful without knowing the following: The MBH radiation is small, may be detected only over a short distance, does not have specific frequency and has a variable long wavelength.

Discussion

Our equations are based upon the assumption $\lambda/\lambda_o = r/r_o$. If $\lambda/\lambda_o = (r/r_o)^{0.5}$ or other relation, the all above equations may be easy modified. The Hawking article was published 34 years ago (1974)[1] and since then hundreds of scientific works based in Hawking work appears and no known facts creates doubt in the possibility of Hawking radiation but neither it is not proven so that the Hawking radiation may not exist. The Large Hadron Collider has the main purpose to create the MBHs and detect the Hawking radiation [5].

Conclusion

The AB- Generator could create a revolution in many industries (electricity, car, ship, transportation, etc.) that allows designing photon rockets and flight to other star systems. The maximum possible efficiency is obtained and a full solution possible for the energy problem of humanity. These overwhelming prospects urge us to research and develop this achievement of science.

References:

(The reader may find some of related articles at the author's web page http://Bolonkin.narod.ru/p65.htm; http://arxiv.org , http://www.scribd.com search "Bolonkin"; http://aiaa.org search "Bolonkin"; and in the author's books: "*Non-Rocket Space Launch and Flight*", Elsevier, London, 2006, 488 pages http://www.scribd.com/doc/24056182; "*New Concepts, Ideas, Innovations in Aerospace, Technology and Human Science*", NOVA, 2007, 502 pages, http://www.scribd.com/doc/24057071 ; "*Macro-Projects: Environment and Technology*", NOVA 2008, 536 pages, http://www.scribd.com/doc/24057930; and "New Technologies and Revolutionary Projects", Scribd, 2010, 324 pgs, http://www.scribd.com/doc/32744477).

1. Hawking, S.W. (1974), "Black hole explosions?", *Nature* 248: 30–31, doi:10.1038/248030a0, http://www.nature.com/nature/journal/v248/n5443/abs/248030a0.html.

2. Bolonkin A.A., Non-Rocket Space Launch and Flight, Elsevier, 2006, 488 pgs. http://www.archive.org/details/Non-rocketSpaceLaunchAndFlight , http://www.scribd.com/doc/24056182

3. Bolonkin A.A., "New Concepts, Ideas, Innovations in Aerospace, Technology and the Human Sciences", NOVA, 2007, 502 pgs. http://www.scribd.com/doc/24057071 , http://www.archive.org/details/NewConceptsIfeasAndInnovationsInAerospaceTechnologyAndHumanSciences

4. Bolonkin A.A., Cathcart R., "Macro-Projects: Environments and Technologies", NOVA, 2008, 536 pgs. , http://www.archive.org/details/Macro-projectsEnvironmentsAndTechnologies , http://www.scribd.com/doc/24057930

5. Bolonkin A.A., "New Technologies and Revolutionary Projects", Sbcribd, 2010, 324 pgs, http://www.archive.org/details/NewTechnologiesAndRevolutionaryProjects, http://www.scribd.com/doc/32744477

6. Bolonkin A.A., Femtotechnology: Nuclear AB-Matter with fantastic properties. http://www.scribd.com/doc/24046679/ , http://www.scipub.org/fulltext/ajeas/ajeas22501-514.pdf

   American Journal of Engineering and Applied Science, Vol. 2, #2, 2009, pp.501-514.

7. Bolonkin A.A., Converting of Matter to Nuclear Energy by AB-Generator. http://www.scribd.com/doc/24048466/ ,

   American Journal of Engineering and Applied Science, Vol. 2, #4, 2009, pp.683-693. http://www.scipub.org/fulltext/ajeas/ajeas24683-693.pdf .

8. Wikipedia. Some background material in this article is gathered from Wikipedia under the Creative Commons license. http://wikipedia.org .

23 April 2009.

Article Criterion for Solar Detonation 1 27 10

Chapter 6

Artificial Explosion of Sun
and AB-Criterion for Solar Detonation

Abstract

The Sun contains ~74% hydrogen by weight. The isotope hydrogen-1 (99.985% of hydrogen in nature) is a usable fuel for fusion thermonuclear reactions.

This reaction runs slowly within the Sun because its temperature is low (relative to the needs of nuclear reactions). If we create higher temperature and density in a limited region of the solar interior, we may be able to produce self-supporting detonation thermonuclear reactions that spread to the full solar volume. This is analogous to the triggering mechanisms in a thermonuclear bomb. Conditions within the bomb can be optimized in a small area to initiate ignition, then spread to a larger area, allowing producing a hydrogen bomb of any power. In the case of the Sun certain targeting practices may greatly increase the chances of an artificial explosion of the Sun. This explosion would annihilate the Earth and the Solar System, as we know them today.

The reader naturally asks: Why even contemplate such a horrible scenario? It is necessary because as thermonuclear and space technology spreads to even the least powerful nations in the centuries ahead, a dying dictator having thermonuclear missile weapons can produce (with some considerable mobilization of his military/industrial complex)— an artificial explosion of the Sun and take into his grave the whole of humanity. It might take tens of thousands of people to make and launch the hardware, but only a very few need know the final targeting data of what might be otherwise a weapon purely thought of (within the dictator's defense industry) as being built for peaceful, deterrent use.

Those concerned about Man's future must know about this possibility and create some protective system—or ascertain on theoretical grounds that it is entirely impossible.

Humanity has fears, justified to greater or lesser degrees, about asteroids, warming of Earthly climate, extinctions, etc. which have very small probability. But all these would leave survivors --nobody thinks that the terrible annihilation of the Solar System would leave a single person alive. That explosion appears possible at the present time. In this paper is derived the 'AB-Criterion' which shows conditions wherein the artificial explosion of Sun is possible. The author urges detailed investigation and proving or disproving of this rather horrifying possibility, so that it may be dismissed from mind—or defended against.

Key words: Artificial explosion of Sun, annihilation of solar system, criterion of nuclear detonation, nuclear detonation wave, detonate Sun, artificial supernova.
* This work is written together J. Friedlander. He corrected the author's English, wrote together with author Abstract, Sections 8, 10 ("Penetration into Sun" and "Results"), and wrote Section 11 "Discussion" as the solo author.

1. Introduction

Information about Sun. The Sun is the star at the center of the Solar System. The Earth and other matter (including other planets, asteroids, meteoroids, comets and dust) orbit the Sun, which by itself accounts for about 99.8% of the solar system's mass. Energy from the Sun—in the form of sunlight—

supports almost all life on Earth via photosynthesis, and drives the Earth's climate and weather.

The Sun is composed of hydrogen (about 74% of its mass, or 92% of its volume), helium (about 25% of mass, 7% of volume), and trace quantities of other elements. The Sun has a spectral class of G2V. *G2* implies that it has a surface temperature of approximately 5,500 K (or approximately 9,600 degrees Fahrenheit / 5,315 Celsius).

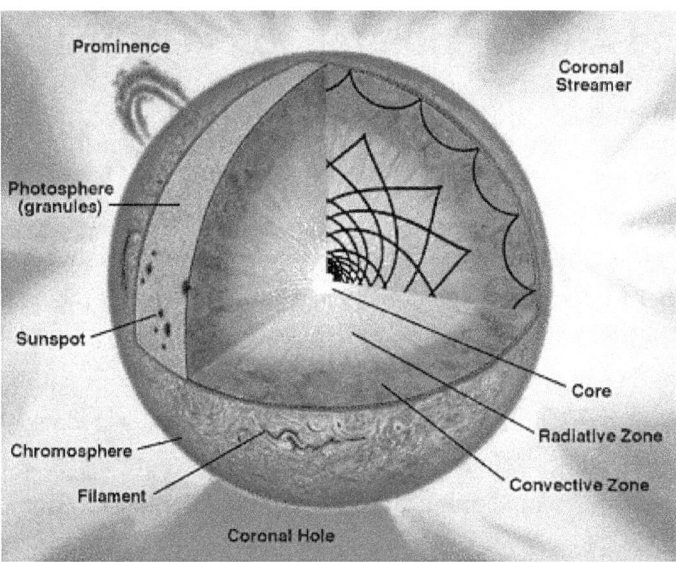

Fig.1. Structure of Sun

Sunlight is the main source of energy to the surface of Earth. The solar constant is the amount of power that the Sun deposits per unit area that is directly exposed to sunlight. The solar constant is equal to approximately 1,370 watts per square meter of area at a distance of one AU from the Sun (that is, on or near Earth). Sunlight on the surface of Earth is attenuated by the Earth's atmosphere so that less power arrives at the surface—closer to 1,000 watts per directly exposed square meter in clear conditions when the Sun is near the zenith.

The Sun is about halfway through its main-sequence evolution, during which nuclear fusion reactions in its core fuse hydrogen into helium. Each second, more than 4 million tonnes of matter are converted into energy within the Sun's core, producing neutrinos and solar radiation; at this rate, the sun will have so far converted around 100 earth-masses of matter into energy. The Sun will spend a total of approximately 10 billion years as a main sequence star.

The core of the Sun is considered to extend from the center to about 0.2 solar radii. It has a density of up to 150,000 kg/m^3 (150 times the density of water on Earth) and a temperature of close to 13,600,000 kelvins (by contrast, the surface of the Sun is close to 5,785 kelvins (1/2350th of the core)). Through most of the Sun's life, energy is produced by nuclear fusion through a series of steps called the *p-p* (proton-proton) chain; this process converts hydrogen into helium. The core is the only location in the Sun that produces an appreciable amount of heat via fusion: the rest of the star is heated by energy that is transferred outward from the core. All of the energy produced by fusion in the core must travel through many successive layers to the solar photosphere before it escapes into space as sunlight or kinetic energy of particles.

About 3.4×10^{38} protons (hydrogen nuclei) are converted into helium nuclei every second (out of about $\sim 8.9 \times 10^{56}$ total amount of free protons in Sun), releasing energy at the matter-energy conversion rate of 4.26 million tonnes per second, 383 yottawatts (383×10^{24} W) or 9.15×10^{10} megatons of TNT per second. This corresponds to extremely low rate of energy production in the Sun's core - about 0.3 µW/cm^3, or about 6 µW/kg. For comparison, an ordinary candle produces heat at the rate 1 W/cm^3, and human body - at the rate of 1.2 W/kg. Use of plasma with similar parameters as solar interior plasma for energy production on Earth is completely impractical - as even a modest 1 GW fusion power plant would require about 170 billion tonnes of plasma occupying almost one cubic mile. Thus all terrestrial fusion reactors require much higher plasma temperatures than those in Sun's interior to be viable.

The rate of nuclear fusion depends strongly on density (and particularly on temperature), so the fusion rate in the core is in a self-correcting equilibrium: a slightly higher rate of fusion would cause the core to heat up more and expand slightly against the weight of the outer layers, reducing the fusion rate and correcting the perturbation; and a slightly lower rate would cause the core to cool and shrink slightly, increasing the fusion rate and again reverting it to its present level.

The high-energy photons (gamma and X-rays) released in fusion reactions are absorbed in only few millimeters of solar plasma and then re-emitted again in random direction (and at slightly lower energy) - so it takes a long time for radiation to reach the Sun's surface. Estimates of the "photon travel time" range from as much as 50 million years to as little as 17,000 years. After a final trip through the convective outer layer to the transparent "surface" of the photosphere, the photons escape as visible light. Each gamma ray in the Sun's core is converted into several million visible light photons before escaping into space. Neutrinos are also released by the fusion reactions in the core, but unlike photons they very rarely interact with matter, so almost all are able to escape the Sun immediately.

This reaction is very slowly because the solar temperatute is very lower of Coulomb barrier.

The Sun's current age, determined using computer models of stellar evolution and nucleocosmochronology, is thought to be about 4.57 billion years.

Astronomers estimate that there are at least 70 sextillion (7×10^{22}) stars in the observable universe. That is 230 billion times as many as the 300 billion in the Milky Way.

Atmosphere of Sun. The parts of the Sun above the photosphere are referred to collectively as the *solar atmosphere*. They can be viewed with telescopes operating across the electromagnetic spectrum, from radio through visible light to gamma rays, and comprise five principal zones: the *temperature minimum*, the chromosphere, the transition region, the corona, and the heliosphere.

The chromosphere, transition region, and corona are much hotter than the surface of the Sun; the reason why is not yet known. But their density is low.

The coolest layer of the Sun is a temperature minimum region about 500 km above the photosphere, with a temperature of about 4,000 K.

Above the temperature minimum layer is a thin layer about 2,000 km thick, dominated by a spectrum of emission and absorption lines. It is called the *chromosphere* from the Greek root *chroma*, meaning color, because the chromosphere is visible as a colored flash at the beginning and end of total eclipses of the Sun. The temperature in the chromosphere increases gradually with altitude, ranging up to around 100,000 K near the top.

Above the chromosphere is a transition region in which the temperature rises rapidly from around 100,000 K to coronal temperatures closer to one million K. The increase is because of a phase transition as helium within the region becomes fully ionized by the high temperatures. The transition region does not occur at a well-defined altitude. Rather, it forms a kind of nimbus around chromospheric features such as spicules and filaments, and is in constant, chaotic motion. The transition region is not easily visible from Earth's surface, but is readily observable from space by instruments sensitive to the far ultraviolet portion of the spectrum.

The corona is the extended outer atmosphere of the Sun, which is much larger in volume than the Sun itself. The corona merges smoothly with the solar wind that fills the solar system and heliosphere. The low corona, which is very near the surface of the Sun, has a particle density of 10^{14} m^{-3}–10^{16} m^{-3}. (Earth's atmosphere near sea level has a particle density of about 2×10^{25} m^{-3}.) The temperature of the corona is several million kelvin. While no complete theory yet exists to account for the temperature of the corona, at least some of its heat is known to be from magnetic reconnection.

Physical characteristics of Sun: Mean diameter is 1.392×10^6 km (109 Earths). Volume is 1.41×10^{18} km³ (1,300,000 Earths). Mass is 1.988 435×10³⁰ kg (332,946 Earths). Average density is 1,408 kg/m³. Surface temperature is 5785 K (0.5 eV). Temperature of corona is 5 MK (0.43 keV). Core temperature is ~13.6 MK (1.18 keV). Sun radius is $R = 696 \times 10^3$ km, solar gravity $g_c = 274$ m/s². Photospheric composition of Sun (by mass): Hydrogen 73.46 %; Helium 24.85 %; Oxygen 0.77 %; Carbon 0.29 %; Iron 0.16 %; Sulphur 0.12 %; Neon 0.12 %; Nitrogen 0.09 %; Silicon 0.07 %; Magnesium 0.05 %.

Sun photosphere has thickness about $7 \times 10^{-4} R$ (490 km) of Sun radius R, average temperature 5.4×10^3 K, and average density 2×10^{-7} g/cm³ ($n = 1.2 \times 10^{23}$ m⁻³). Sun convection zone has thickness about $0.15 R$, average temperature 0.25×10^6 K, and average density 5×10^{-7} g/cm³. Sun intermediate (radiation) zone has thickness about $0.6 R$, average temperature 4×10^6 K, and average density 10 g/cm³. Sun core has thickness about $0.25 R$, average temperature 11×10^6 K, and average density 89 g/cm³.

Detonation is a process of combustion in which a supersonic shock wave is propagated through a fluid due to an energy release in a reaction zone. This self-sustained detonation wave is different from a deflagration, which propagates at a subsonic rate (i.e., slower than the sound speed in the material itself).

Detonations can be produced by explosives, reactive gaseous mixtures, certain dusts and aerosols.

The simplest theory to predict the behavior of detonations in gases is known as Chapman-Jouguet (CJ) theory, developed around the turn of the 20th century. This theory, described by a relatively simple set of algebraic equations, models the detonation as a propagating shock wave accompanied by exothermic heat release. Such a theory confines the chemistry and diffusive transport processes to an infinitely thin zone.

A more complex theory was advanced during World War II independently by Zel'dovich, von Neumann, and Doering. This theory, now known as ZND theory, admits finite-rate chemical reactions and thus describes a detonation as an infinitely thin shock wave followed by a zone of exothermic chemical reaction. In the reference frame in which the shock is stationary, the flow following the shock is subsonic. Because of this, energy release behind the shock is able to be transported acoustically to the shock for its support. For a self-propagating detonation, the shock relaxes to a speed given by the Chapman-Jouguet condition, which induces the material at the end of the reaction zone to have a locally sonic speed in the reference frame in which the shock is stationary. In effect, all of the chemical energy is harnessed to propagate the shock wave forward.

Both CJ and ZND theories are one-dimensional and steady. However, in the 1960s experiments revealed that gas-phase detonations were most often characterized by unsteady, three-dimensional structures, which can only in an averaged sense be predicted by one-dimensional steady theories. Modern computations are presently making progress in predicting these complex flow fields. Many features can be qualitatively predicted, but the multi-scale nature of the problem makes detailed quantitative predictions very difficult.

2. Statement of Problem, Main Idea and Our Aim

The present solar temperature is far lower than needed for propagating a runaway thermonuclear reaction. In Sun core the temperature is only ~13.6 MK (0.0012 MeV). The Coulomb barrier for protons (hydrogen) is more then 0.4 MeV. Only very small proportions of core protons take part in the thermonuclear reaction (they use a tunnelling effect). Their energy is in balance with energy emitted by Sun for the Sun surface temperature 5785 K (0.5 eV).

We want to clarify: If we create a zone of limited size with a high temperature capable of overcoming the Coulomb barrier (for example by insertion of a thermonuclear warhead) into the solar photosphere (or lower), can this zone ignite the Sun's photosphere (ignite the Sun's full load of thermonuclear fuel)? Can this zone self-support progressive runaway reaction propagation for a significant proportion of the available thermonuclear fuel?

If it is possible, researchers can investigate the problems: What will be the new solar temperature? Will this be metastable, decay or runaway? How long will the transformed Sun live, if only a minor change? What the conditions will be on the Earth?

Why is this needed?

As thermonuclear and space technology spreads to even the least powerful nations in the decades and centuries ahead, a dying dictator having thermonuclear weapons and space launchers can produce (with some considerable mobilization of his military/industrial complex)— the artificial explosion of the Sun and take into his grave the whole of humanity.

It might take tens of thousands of people to make and launch the hardware, but only a very few need know the final targeting data of what might be otherwise a weapon purely thought of (within the dictator's defense industry) as being built for peaceful, 'business as usual' deterrent use. Given the hideous history of dictators in the twentieth century and their ability to kill technicians who had outlived their use (as well as major sections of entire populations also no longer deemed useful) we may assume that such ruthlessness is possible.

Given the spread of suicide warfare and self-immolation as a desired value in many states, (in several cultures—think Berlin or Tokyo 1945, New York 2001, Tamil regions of Sri Lanka 2006) what might obtain a century hence? All that is needed is a supportive, obedient defense complex, a 'romantic' conception of mass death as an ideal—even a religious ideal—and the realization that his own days at power are at a likely end. It might even be launched as a trump card in some (to us) crazy internal power struggle, and plunged into the Sun and detonated in a mood of spite by the losing side. *'Burn baby burn'!*

A small increase of the average Earth's temperature over 0.4 K in the course of a century created a panic in humanity over the future temperature of the Earth, resulting in the Kyoto Protocol. Some stars with active thermonuclear reactions have temperatures of up to 30,000 K. If not an explosion but an enchanced burn results the Sun might radically increase in luminosity for –say--a few hundred years. This would suffice for an average Earth temperature of hundreds of degrees over 0 C. The oceans would evaporate and Earth would bake in a Venus like greenhouse, or even lose its' atmosphere entirely.

Thus we must study this problem to find methods of defense from human induced Armageddon.

The interested reader may find needed information in [1]-[4].

3. Theory and estimations

1. **Coulomb barrier (repulsion).** Energy is needed for thermonuclear reaction may be computed by equations

$$E = \frac{kZ_1Z_2e^2}{r} = 2.3 \cdot 10^{-28} \frac{Z_1Z_2}{r} \ [J] \quad \text{or} \quad E = \frac{kZ_1Z_2e}{r} = 1.44 \cdot 10^{-9} \frac{Z_1Z_2}{r} \ [eV], \qquad (1)$$

$$\text{where} \quad r = r_1 + r_2, \quad r_i = (1.2 \div 1.5) \times 10^{-15} \sqrt[3]{A_i}$$

where E is energy needed for forcing contact between two nuclei, J or eV; $k = 9\times10^9$ is electrostatic constant, Nm^2/C^2; Z is charge state; $e = 1.6\times10^{-19}$ is charge of proton, C; r is distance between nucleus centers, m; r_i is radius of nucleus, m; $A = Z + N$ is nuclei number, N is number neutrons into given (i = 1, 2) nucleus.

The computations of average temperature (energy) for some nucleus are presented in Table #1 below. We assume that the first nucleus is moving; the second (target) nucleus is motionless.

Table 1. Columb barrier of some nuclei pairs.

Reaction	E, MeV	Reaction	E, MeV	Reaction	E, MeV	Reaction	E, MeV
p + p	0.53	T+p	0.44	^{6}L+p	1.13	^{13}C+p	1.9
D + p	0.47	D+d	0.42	^{7}Be+p	1.5	^{12}C+^{4}He	3.24

In reality the temperature of plasma may be significantly lower than in table 1 because the nuclei have different velocity. Parts of them have higher velocity (see Maxwell distribution of nuclei speed in plasma), some of the nuclei do not (their energy are summarized), and there are tunnel effects. If the temperature is significantly lower, then only a small part of the nuclei took part in reaction and the fuel burns very slowly. This case we have--happily in the present day Sun where the temperature in core has only 0.0012 MeV and the Sun can burn at this rate for billions of years.

The ratio between temperatures in eV and in K is

$$T_K = 1.16 \times 10^4 T_e, \quad T_e = 0.86 \times 10^{-4} T_K. \qquad (2)$$

2. The energy of a nuclear reaction. The energy and momentum conservation laws define the energetic relationships for a nuclear reaction [1]-[2].

When a reaction $A(a,b)B$ occurs, the quantity

$$Q = [(M_A + M_a) - (M_B + M_b)]c^2, \qquad (3)$$

where M_i are the masses of the particles participating in the reaction and c is the speed of light, Q is the reaction energy.

Usually *mass defects* ΔM are used, instead of masses, for computing Q:

$$Q = (\Delta M_A + \Delta M_a) - (\Delta M_B + \Delta M_b). \qquad (4)$$

The mass defect is the quantity $\Delta M = M - A$ where M is the actual mass of the particle (atom), A is the so-called mass number, i.e. the total number of nucleons (protons and neutrons) in the atomic nucleus. If M is expressed in atomic mass units (a.m.u.) and A is assigned the same unit, then ΔM is also expressed in a.m.u. One a.m.u. represent 1/12 of the ^{12}C nuclide mass and equals 1.6605655×10^{-27} kg. For calculations of reaction energies it is more convenient to express ΔM in kilo-electronvolts: a.m.u. = 931501.59 keV.

Employing the mass defects, one can handle numbers that are many times smaller than the nuclear masses or the binding energies.

Kinetic energy may be released during the course of a reaction (exothermic reaction) or kinetic energy may have to be supplied for the reaction to take place (endothermic reaction). This can be calculated by reference to a table of very accurate particle rest masses (see http://physics.nist.gov/PhysRefData/Compositions/index.html). The reaction energy (the "Q-value") is positive for exothermal reactions and negative for endothermal reactions.

The other method calculate of thermonuclear energy is in [1]. For a nucleus of atomic number Z, mass number A, and Atomic mass M(Z,A), the binding energy is

$$Q = [ZM(^1H) + (A-Z)m_n - M(Z,A)]c^2, \qquad (5)$$

where $M(^1H)$ is mass of a hydrogen atom and m_n is mass of neutron. This equation neglects a small correction due to the binding energy of the atomic electrons.

The binding energy per nucleus Q/A, varies only slightly in the range of 7 - 9 MeV for nuclei with A > 12.

The binding energy can be approximately calculated from Weizsacker's semiempirical formula:

$$Q = a_v A - a_s A^{2/3} - a_c Z(Z-1)A^{-1/3} - a_{sym}(A-2Z)^2/A + \delta, \qquad (6)$$

where δ accounts for pairing of like nucleons and has the value $+a_p A^{-3/4}$ for Z and N both even, $-a_p A^{-3/4}$ for Z and N both odd, and zero otherwise (A odd). The constants in this formula must be adjusted for the best agreement with data: typical value are a_v = 15.5 MeV, a_s = 16.8 MeV, a_c = 0.72 MeV, a_{sym} = 23 MeV, and a_p = 34 MeV.

The binding energy per nucleon of the helium-4 nucleus is unusually high, because the He-4 nucleus is doubly magic. (The He-4 nucleus is unusually stable and tightly-bound for the same reason that the helium atom is inert: each pair of protons and neutrons in He-4 occupies a filled **1s** nuclear orbital in the same way that the pair of electrons in the helium atom occupies a filled **1s** electron orbital). Consequently, alpha particles appear frequently on the right hand side of nuclear reactions.

The energy released in a nuclear reaction can appear mainly in one of three ways:

- kinetic energy of the product particles
- emission of very high energy photons, called gamma rays
- some energy may remain in the nucleus, as a metastable energy level.

When the product nucleus is metastable, this is indicated by placing an asterisk ("*") next to its atomic number. This energy is eventually released through nuclear decay.

If the reaction equation is balanced, that does not mean that the reaction really occurs. The rate at which reactions occur depends on the particle energy, the particle flux and the reaction cross section.

In the initial collision which begins the reaction, the particles must approach closely enough so that the short range strong force can affect them. As most common nuclear particles are positively charged, this means they must overcome considerable electrostatic repulsion before the reaction can begin. Even if the target nucleus is part of a neutral atom, the other particle must penetrate well beyond the electron cloud and closely approach the nucleus, which is positively charged. Thus, such particles must be first accelerated to high energy, for example by very high temperatures, on the order of millions of degrees, producing thermonuclear reactions

Also, since the force of repulsion is proportional to the product of the two charges, reactions between heavy nuclei are rarer, and require higher initiating energy, than those between a heavy and light nucleus; while reactions between two light nuclei are commoner still.

Neutrons, on the other hand, have no electric charge to cause repulsion, and are able to effect a nuclear reaction at very low energies. In fact at extremely low particle energies (corresponding, say, to thermal equilibrium at room temperature), the neutron's de Broglie wavelength is greatly increased, possibly greatly increasing its capture cross section, at energies close to resonances of the nuclei involved. Thus low energy neutrons *may* be even more reactive than high energy neutrons.

Table 2. Exothermic thermonuclear reactions.

№	Reaction	Energy of reaction MeV	σ_{max} barn $E \leq 1$ MeV	E of σ_{max} MeV	№	Reaction MeV	Energy of reaction MeV	σ_{max} barn $E \leq 1$ MeV	E of σ_{max} MeV
1	p+p→d+e$^+$+ν	2.2	10^{-23}	-	15	d+^{6}Li→^{7}Li +p	5.0	0.01	1
2	p+d→^{3}He+γ	5.5	10^{-6}	-	16	d+^{6}Li→2^{4}He	22.4	0.026	0.60
3	p+t→^{4}He+γ	19.7	10^{-6}	-	17	d+^{7}Li→2^{4}He+n	15.0	10^{-3}	0.2
4	d+d→t+p	4.0	0.16	2	18	p+^{9}Be→2^{4}He+d	0.56	0.46	0.33
5	d+d→^{3}He+n	3.3	0.09	1	19	p+^{9}Be→^{6}Li+^{4}He	2.1	0.34	0.33
6	d+d→^{4}He+γ	24	-	-	20	p+^{11}B→3^{4}He	8.7	0.6	0.675
7	d+t→^{4}He+n	17.6	5	0.13	21	p+^{15}N→^{12}C+^{4}He	5.0	0.6	1.2
8	t+d→^{4}He+n	17.6	5	0.195	22	d+^{6}Li→^{7}Be+n	3.4	0.01	0.3
9	t+t→^{4}He+2n	11.3	0.1	1	23	^{3}He+t→^{4}He+d	14.31	0.7	≈1
10	d+^{3}He→^{4}He+p	18.4	0.71	0.47	24	^{3}H+^{4}He→^{7}Li+γ	2.457	$7 \cdot 10^{-5}$	≈3
11	^{3}He+^{3}He→^{4}He+2p	12.8	-	-	25	^{3}H+d→^{4}He	17.59	$5 \cdot 10^{-4}$	≈2
12	n+^{6}Li→^{4}He+t	4,8	2.6	0.26	26	^{12}C+p→^{13}N+γ	1.944	10^{-6}	0.46
13	p+^{6}Li→^{4}He+^{3}He	4,0	10^{-4}	0.3	27	^{13}C+p→^{14}N+γ	7.55	10^{-4}	0.555
14	p+^{7}Li→2^{4}He+γ	17.3	$6 \cdot 10^{-3}$	0.44	28	^{3}He+^{4}He→^{7}Be+γ	1.587	10^{-6}	≈8

Here are: p (or ¹H) - proton, d (or D, or ²H) - deuterium, t (or T, or ³H) - tritium, n - neutron, He - helium, Li - lithium, Be - beryllium, B - barium, C - carbon, N - hydrogen, ν - neutrino, γ - gamma radiation.

3. Distribution of thermonuclear energy between particles. In most cases the result of thermonuclear reaction is more than one product. As you see in Table 2 that may be "He" and neutron or proton. The thermonuclear energy distributes between them in the following manner:

$$\text{From} \quad E = E_1 + E_2 = \frac{m_1 V_1^2}{2} + \frac{m_2 V_2^2}{2}, \quad m_1 V_1 = m_2 V_2,$$
$$\text{we have} \quad \frac{E_1}{E} = \frac{m_2}{m_1 + m_2} = \frac{\mu_2}{\mu_1 + \mu_2}, \quad E_2 = E - E_1 \tag{7}$$

where m is particle mass, kg; V is particle speed, m/s; E is particle energy, J; $\mu = m_i/m_p$ is relative particle mass. Lower indexes "$_{1, 2}$" are number of particles.

After some collisions the energy $E = kT$ (temperature) of different particles may be closed to equal.

4. The power density produced in thermonuclear reaction may be computed by the equation

$$P = E n_1 n_2 <\sigma v>, \tag{8}$$

where E is energy of single reaction, eV or J; n_1 is density (number particles in cm³) the first component; n_2 is density (number particles in cm³) the second component; $<\sigma v>$ is reaction rate, in cm³/s; σ is cross section of reaction, cm², 1 barn = 10^{-24} cm²; v is speed of the first component, cm/s; P is power density, eV/cm³ or J/cm³. Cross section of reaction before σ_{max} very strongly depends from temperature and it is obtainable by experiment. They can have the maximum resonance. For very high temperatures the σ may be close to the nuclear diameter.

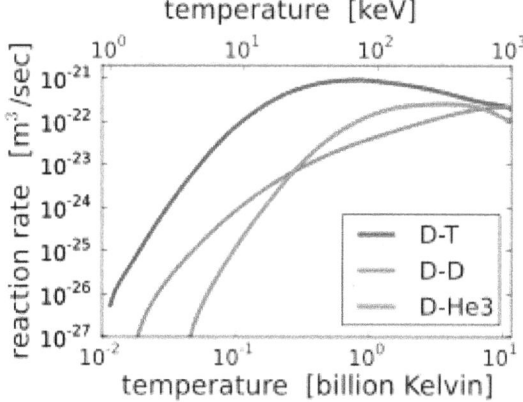

Fig.2. Reaction rate $<\sigma v>$ via plasma temperature for D-T (top), D-D (middle) and D-³He (bottom in left side).

The terminal velocity of the reaction components (electron and ions) are

$$v_{Te} = (kT_e / m_e)^{1/2} = 4.19 \times 10^7 T_e^{1/2}. \quad cm/s, \tag{9}$$

$$v_{Ti} = (kT_i/m_i)^{1/2} = 9.79 \times 10^7 (T_i/\mu_i)^{1/2}. \quad cm/s, \qquad (10)$$

where T is temperature in eV; $\mu_i = m_i/m_p$ is ratio of ion mass to proton mass.

The sound velocity of ions is

$$v = \left(\frac{\gamma z k T_k}{m_i}\right)^{1/2}, \qquad (11)$$

where $\gamma \approx (1.2 \div 1.4)$ is adiabatic coefficient; z is number of charge ($z = 1$ for p), T_k is plasma temperature in K; m_i is mass of ion.

The deep of penetration of outer radiation into plasma is

$$d = 5.31 \cdot 10^5 n_e^{-1/3}, \quad [cm]$$

where n_e is number of electrons in unit of volume.

In internal plasma detonation there is no loss in radiation because the plasma reflects the radiation.

4. Possible Thermonuclear Reactions to Power a Hypothetical Solar Explosion

The Sun mass is ~74% hydrogen and 25% helium. Possibilities exist for the following self-supporting nuclear reactions in the hydrogen medium: proton chain reaction, CNO cycle, Triple-alpha process, Carbon burning process, Neon burning process, Oxygen burning process, Silicon burning process.

For our case of particular interest (a most probable candidate) the proton-proton chain reaction. It is more exactly the reaction $p + p$.

The proton-proton chain reaction is one of several fusion reactions by which stars convert hydrogen to helium, the primary alternative being the CNO cycle. The proton-proton chain dominates in stars the size of the Sun or less.

The first step involves the fusion of two hydrogen nuclei ^{1}H (protons) into deuterium ^{2}H, releasing a positron and a neutrino as one proton changes into a neutron.

$$^1H + {}^1H \rightarrow {}^2H + \underline{e^+} + \underline{\nu_e}. \qquad (12)$$

with the neutrinos released in this step carrying energies up to 0.42 MeV.

The positron immediately annihilates with an electron, and their mass energy is carried off by two gamma ray photons.

$$e^+ + e^- \rightarrow 2\underline{\gamma} + 1.02 \text{ MeV}. \qquad (13)$$

After this, the deuterium produced in the first stage can fuse with another hydrogen to produce a light isotope of helium, ^{3}He:

$$^2H + {}^1H \rightarrow {}^3He + \gamma + 5.49 \text{ MeV}. \qquad (14)$$

From here there are three possible paths to generate helium isotope ^{4}He. In pp1 helium-4 comes from fusing two of the helium-3 nuclei produced; the pp2 and pp3 branches fuse ^{3}He with a pre-existing ^{4}He to make Beryllium-7. In the Sun, branch pp1 takes place with a frequency of 86%, pp2 with 14% and pp3 with 0.11%. There is also an extremely rare pp4 branch.

The pp I branch

$$^3\text{He} + {}^3\text{He} \rightarrow {}^4\text{He} + {}^1\text{H} + {}^1\text{H} + 12.86 \text{ MeV}$$

The complete pp I chain reaction releases a net energy of 26.7 MeV. The pp I branch is dominant at temperatures of 10 to 14 megakelvins (MK). Below 10 MK, the PP chain does not produce much ^{4}He.

The pp II branch

$$^3\text{He} + {}^4\text{He} \rightarrow {}^7\text{Be} + \gamma$$

$$^7\text{Be} + e^- \rightarrow {}^7\text{Li} + \nu_e$$

$$^7\text{Li} + {}^1\text{H} \rightarrow {}^4\text{He} + {}^4\text{He}$$

The pp II branch is dominant at temperatures of 14 to 23 MK. 90% of the neutrinos produced in the reaction ^{7}Be(e^-,ν_e)^{7}Li* carry an energy of 0.861 MeV, while the remaining 10% carry 0.383 MeV (depending on whether lithium-7 is in the ground state or an excited state, respectively).

The pp III branch

$$^3\text{He} + {}^4\text{He} \rightarrow {}^7\text{Be} + \gamma$$

$$^7\text{Be} + {}^1\text{H} \rightarrow {}^8\text{B} + \gamma$$

$$^8\text{B} \rightarrow {}^8\text{Be} + e^+ + \nu_e$$

$$^8\text{Be} \leftrightarrow {}^4\text{He} + {}^4\text{He}$$

The pp III chain is dominant if the temperature exceeds 23 MK.

The pp III chain is not a major source of energy in the Sun (only 0.11%), but was very important in the solar neutrino problem because it generates very high energy neutrinos (up to 14.06 MeV).

The pp IV or Hep

This reaction is predicted but has never been observed due to its great rarity (about 0.3 parts per million in the Sun). In this reaction, Helium-3 reacts directly with a proton to give helium-4, with an even higher possible neutrino energy (up to 18.8 MeV).

$$^3\text{He} + {}^1\text{H} \rightarrow {}^4\text{He} + \nu_e + e^+$$

Energy release.

Comparing the mass of the final helium-4 atom with the masses of the four protons reveals that 0.007 or 0.7% of the mass of the original protons has been lost. This mass has been converted into energy, in the form of gamma rays and neutrinos released during each of the individual reactions.

The total energy we get in one whole chain is

$$4{}^1H \rightarrow {}^4He + 26.73 \text{ MeV}.$$

Only energy released as gamma rays will interact with electrons and protons and heat the interior of the Sun. This heating supports the Sun and prevents it from collapsing under its own weight. Neutrinos do not interact significantly with matter and do not help support the Sun against gravitational collapse. The neutrinos in the ppI, ppII and ppIII chains carry away the 2.0%, 4.0% and 28.3% of the energy respectively.

This creates a situation in which stellar nucleosynthesis produces large amounts of carbon and oxygen but only a small fraction of these elements is converted into neon and heavier elements. Both oxygen and carbon make up the *ash* of helium burning. Those nuclear resonances sensitively are arranged to create large amounts of carbon and oxygen, has been controversially cited as evidence of the anthropic principle.

About 34% of this energy is carried away by neutrinos. That reaction is part of solar reaction, but if initial temperature is high, the reaction becomes an explosion.

The detonation wave works a short time. That supports the reactions (12) – (13). They produce energy up to 1.44 MeV. The reactions (12) – (14) produce energy up to 5.8 MeV. But after detonation wave and the full range of reactions the temperature of plasma is more than the temperature needed to pass the Coulomb barrier and the energy of explosion increases by 20 times.

5. Detonation theory

The one dimensional detonation wave may be computed by equations (see Fig.2):

1) Law of mass

$$\frac{D}{V_1} = \frac{v}{V_3}, \qquad (15)$$

where D – speed of detonation, m/s; v – speed of ion sound, m/s about the front of detonation wave (eq.(11)); V_1, V_3 specific density of plasma in points 1, 3 respectively, kg/m³.

2) Law of momentum

$$p_1 + \frac{D^2}{V_1} = p_3 + \frac{v^2}{V_3}, \qquad (16)$$

where p_1, p_3 are pressures, N/m², in point 1, 3 respectively.

3) Law of energy

$$E_3 - E_1 = Q + 0.5(p_3 + p_1)(V_1 - V_3), \qquad (17)$$

where E_3, E_1 – internal energy, J/kg, of mass unit in point 3, 1 respectively, Q is nuclear energy, J/kg.

4) Speed of detonation is

$$D = \sqrt{2Q(\gamma^2 - 1)}, \quad (18)$$

$\gamma \approx 1.2 \div 1.4$ is adiabatic coefficient.

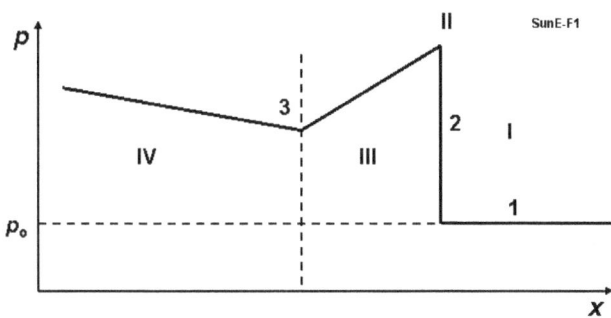

Fig. 3. Pressure in detonation wave. I – plasma, II – front of detonation wave, III – zone of the initial thermonuclear fusion reaction, IV – products of reaction and next reaction, p_o – initial pressure, x – distance.

6. Model of artificial Sun explosion and estimation of ignition

Thermonuclear reactions proceeding in the Sun's core are under high temperature and pressure. However the core temperature is substantially lower than that needed to overcome the Columb barrier. That way the thermonuclear reaction is very slow and the Sun's life cycle is about 10 billion years. But that is enough output to keep the Sun a plasma ball, hot enough for life on Earth to exist. Now we are located in the middle of the Sun's life and have about 5 billions years until the Sun becomes a Red Giant.

However, this presumes that the Sun is stable against deliberate tampering. Supposing our postulations are correct, the danger exists that introducing a strong thermonuclear explosion into the Sun which is a container of fuel for thermonuclear reactions, the situation can be cardinally changed. For correct computations it is necessary to have a comprehensive set of full initial data (for example, all cross-section areas of all nuclear reactions) and supercomputer time. The author does not have access to such resources. That way he can only estimate probability of these reactions, their increasing or decreasing. Supportive investigations are welcome in order to restore confidence in humanity's long term future.

7. AB-Criterion for Solar Detonation

A self-supporting detonation wave is possible if the speed of detonation wave is greater or equals the ion sound speed:

$$D \geq v, \quad \text{where} \quad D = \sqrt{2Q(\gamma^2 - 1)}, \quad v = \left(\frac{\gamma z k T_k}{m_i} \right)^{1/2}. \quad (19)$$

Here Q is a nuclear specific heat [J/kg], $\gamma = 1.2 \div 1.4$ is adiabatic coefficient (they are noted in (17)-(18)); z is number of the charge of particle after fusion reaction (z =1 for ^2H), k = 1.36×10^{-23} is Boltzmann constant, J/K; T_k is temperature of plasma after fusion reaction in Kelvin degrees; $m_i = \mu m_p$ is mass of ion after fusion reaction, kg; m_p = 1.67×10^{-27} kg is mass of proton; μ is relative mass, μ =2 for ^2H.

When we have sign ">" the power of the detonation wave increases, when we have the sign "<" it decreases.

Substitute two last equations in the first equation in (19) we get

$$D^2 \geq v^2, \quad 2Q(\gamma^2 - 1) \geq \frac{\gamma z k T_k}{m_i}, \quad \text{where} \quad Q = \frac{feE\tau}{nm_p} = \frac{1}{4} n^2 eE <\sigma v> \frac{\tau}{nm_p}, \quad (20)$$

where f is speed of nuclear reaction, s/m³; $e = 1.6 \times 10^{-19}$ is coefficient for converting the energy from electron-volts to joules; E is energy of reaction in eV; n is number particles (p - protons) in m³; $<\sigma v>$ is reaction rate, m³/s (fig.1), $m_i = 2 m_p$, τ is time, sec.

From (20) we get the AB-Criterion for artificial Sun explosion:

$$n\tau \geq \frac{\gamma z k T_k}{(\gamma^2 - 1)eE <\sigma v>} = \frac{1.16 \cdot 10^4 \gamma z k T_e}{(\gamma^2 - 1)eE <\sigma v>} = \frac{\gamma z T_e}{(\gamma^2 - 1)E <\sigma v>}, \quad (21)$$

where T_e is temperature of plasma after reaction in eV.

The offered AB-Criterion (21) is different from the well-known Lawson criterion

$$n_e \tau_e > \frac{12 k_B T_k}{E_{ch} <\sigma v>},$$

where E_{ch} is energy of reaction in keV, k_B is Boltzmann constant.

The offered AB-Criterion contains the γ adiabatic coefficient and z – number of electric charge in the electron charges. It is not surprising because Lawson derived his criterion from the condition where the energy of the reaction must be greater than the loss of energy by plasma into the reactor walls, where

$$W_{reaction} > W_{loss}.$$

In our case no the reactor walls and plasma reflects the any radiation.

The offered AB-Criterion is received from the condition (19): Speed of self-supporting detonation wave must be greater than the speed of sound where

$$D > v.$$

For main reaction $p + p$ the AB-Criterion (21) has a form

$$n\tau \geq \frac{T_e}{E <\sigma v>}. \quad (21a)$$

Estimation. Let us take the first step of the reaction ¹H + ¹H (12)-(13) having in point 3 (fig.2) $T_e = 10^5$ eV, $E \approx 1.44 \times 10^6$ eV, $<\sigma v> \approx \times 10^{-22}$. Substituting them in equation (21) we receive

$$n\tau > 0.7 \times 10^{21}. \quad (22)$$

The Sun surface (photosphere) has density $n = 10^{23}$ 1/m³, the encounter time of protons in the hypothetical detonation wave III (fig.2) may be over 0.01 sec. The values in left and right sides of (22) have the same order. That means a thermonuclear bomb exploded within the Sun may convceivably be able to serve as a detonator which produces a self-supported nuclear reaction and initiates the artificial explosion of the Sun.

After the initial reaction the temperature of plasma is very high (>1 MeV) and time of next reaction may be very large (hundreds of seconds), the additional energy might in these conditions increase up to 26 MeV.

A more accurate computation is possible but will require cooperation of an interested supercomputer team with the author, or independent investigations with similar interests.

8. Penetration of Thermonuclear Bomb Into Sun

The Sun is a ball of plasma (ionized gases), not a solid body. A properly shielded thermonuclear bomb can permeate deep into the Sun. The warhead may be protected on its' way down by a special high reflectivity mirror offered, among others, by author A.A. Bolonkin in 1983 [12] and described in [7] Chapters 12, 3A, [8] Ch.5, [9]-[12]. This mirror allows to maintain a low temperature of the warhead up to the very boundary of the solar photosphere. At that point its' velocity is gigantic, about 617.6 km/s, assuring a rapid penetration for as far as it goes.

The top solar atmosphere is very rarefied; a milliard (US billion) times less than the Earth's atmosphere. The Sun's photosphere has a density approximately 200 times less than the Earth's atmosphere. Some references give a value of only 0.0000002 gm/cm^3 (.1 millibar) at the photosphere surface. Since present day ICBM warheads can penetrate down (by definition) to the 1 bar level (Earth's surface) and that is by no means the boundary of the feasible, the 10 bar level may be speculated to be near-term achievable. The most difficult entry yet was that of the Galileo atmospheric probe on Dec. 7, 1995 [17]. The Galileo Probe was a 45° sphere-cone that entered Jupiter's atmosphere at 47.4 km/s (atmosphere relative speed at 450 km above the 1 bar reference altitude). The peak deceleration experienced was 230 g (2.3 km/s^2). Peak stagnation point pressure before aeroshell jettison was 9 bars (900 kPa). The peak shock layer temperature was approximately 16000 K (and remember this is into hydrogen (mostly) the solar photosphere is merely 5800 K). Approximately 26% of the Galileo Probe's original entry mass of 338.93 kg was vaporized during the 70 second heat pulse. Total blocked heat flux peaked at approximately 15000 W/cm^2 (hotter than the surface of the Sun).

If the entry vehicle was not optimized for slowdown as the Galileo Probe but for penetration like a modern ICBM warhead, with extra ablatives and a sharper cone half-angle, achievable penetration would be deeper and faster. If 70 seconds atmospheric penetration time could be achieved, (with minimal slowdown) perhaps up to 6 % of the way to the center might be achieved by near term technology.

The outer penetration shield of the warhead may be made from carbon (which is an excellent ablative heat protector). The carbon is also an excellent nuclear catalyst of the nuclear reactions in the CNO solar thermonuclear cycle and may significantly increase the power of the initial explosion.

A century hence, what level of penetration of the solar interior is possible? This depth is unknown to the author, exceeding plausible engineering in the near term. Let us consider a hypothetical point (top of the radiation layer) 30 percent of the way from the surface to the core, at the density of 0.2 g/cm^3 with a temperature of 2,000,000° C. No material substance can withstand such heat—for extended periods.

We may imagine however hypothetical penetration aids, analogous to ICBM techniques of a half century ago. Shock waves bearing the brunt of the encountered heat and forcing it aside, the opacity shielding the penetrator. A form of multiple disposable shock cones may be employed to give the last in line a chance to survive; indeed the destruction of the next to last may arm the trigger.

If the heat isolation shield and multiple penetration aids can protect the bomb at near entry velocity for a hellish *10 minute interval,* (which to many may seem impossible but which cannot be excluded without definitive study—remember we are speaking now of centuries hence, not the near term case above—see reference 14) that means the bomb may reach the depth of 350 thousands kilometers or 0.5R, where $R = 696 \times 10^3$ km is Sun's radius.

The Sun density via relative Sun depth may be estimated by the equation

$$n = n_s e^{20.4\bar{h}}, \quad \text{where} \quad \bar{h} = h/R, \tag{23}$$

where $n_s \approx 10^{23}$ 1/m³ is the plasma density on the photosphere surface; h is deep, km; $R = 696 \times 10^3$ is solar radius, km. At a solar interior depth of $h = 0.5R$ the relative density is greater by *27 thousand times* than on the Sun's surface.

Here the density and temperature are significantly more than on the photosphere's surface. And conditions for the detonation wave and thermonuclear reaction are 'better'—from the point of view of the attacker.

9. Estimation of nuclear bomb needed for Sun explosion

Sound speed into plasma headed up $T = 100$ K million degrees is about

$$v \approx 10^2 T^{0.5} \text{ m/s} = 10^6 \text{ m/s}. \tag{24}$$

Time of nuclear explosion (a full nuclear reaction of bomb) is less $t = 10^{-4}$ sec. Therefore the radius of heated Sun photosphere is about $R = vt = 100$ m, volume V is about

$$V = \frac{4}{3}\pi R^3 \approx 4 \cdot 10^6 \text{ m}^3. \tag{25}$$

Density of Sun photosphere is $p = 2 \times 10^{-4}$ kg/m³. Consequently the mass of the heated photosphere is about $m = pV = 1000$ kg.

The requested power of the nuclear bomb for heating this mass for temperature $T = 10^4$ eV (100 K million degrees) is approximately

$$E = 10^3 \times 10^4 / 1.67 \times 10^{-27} \text{ eV} \approx 0.6 \cdot 10^{34} \text{ eV} \approx 2 \cdot 10^{15} \text{ J} \approx 0.5 \text{ Mt}. \tag{26}$$

The requested power of nuclear bomb is about 0.5 Megatons. The average power of the current thermonuclear bomb is 5 – 10 Mt. That means the current thermonuclear bomb may be used as a fuse of Sun explosion. That estimation needs in a more complex computation by a power computer.

10. Results of research

The Sun contains 73.46 % hydrogen by weight. The isotope hydrogen-1 (99.985% of hydrogen in nature) is usable fuel for a fusion thermonuclear reaction.

The p-p reaction runs slowly within the Sun because its temperature is low (relative to the temperatures of nuclear reactions). If we create higher temperature and density in a limited region of the solar interior, we may be able to produce self-supporting, more rapid detonation thermonuclear reactions that may spread to the full solar volume. This is analogous to the triggering mechanisms in a thermonuclear bomb. Conditions within the bomb can be optimized in a small area to initiate ignition, build a spreading reaction and then feed it into a larger area, allowing producing a 'solar hydrogen bomb' of any power—but not necessarily one whose power can be limited. In the case of the Sun certain targeting practices may greatly increase the chances of an artificial explosion of the entire Sun. This explosion would annihilate the Earth and the Solar System, as we know them today.

Author A.A. Bolonkin has researched this problem and shown that an artificial explosion of Sun cannot be precluded. In the Sun's case this lacks only an initial fuse, which induces the self-supporting detonation wave. This research has shown that a thermonuclear bomb exploded within the solar photosphere surface may be the fuse for an accelerated series of hydrogen fusion reactions.

The temperature and pressure in this solar plasma may achieve a temperature that rises to billions of degrees in which all thermonuclear reactions are accelerated by many thousands of times. This power output would further heat the solar plasma. Further increasing of the plasma temperature would, in the worst case, climax in a solar explosion.

The possibility of initial ignition of the Sun significantly increases if the thermonuclear bomb is exploded under the solar photosphere surface. The incoming bomb has a diving speed near the Sun of about 617 km/sec. Warhead protection to various depths may be feasible –ablative cooling which evaporates and protects the warhead some minutes from the solar temperatures. The deeper the penetration before detonation the temperature and density achieved greatly increase the probability of beginning thermonuclear reactions which can achieve explosive breakout from the current stable solar condition.

Compared to actually penetrating the solar interior, the flight of the bomb to the Sun, (with current technology requiring a gravity assist flyby of Jupiter to cancel the solar orbit velocity) will be easy to shield from both radiation and heating and melting. Numerous authors, including A.A. Bolonkin in works [7]-[12] offered and showed the high reflectivity mirrors which can protect the flight article within the orbit of Mercury down to the solar surface.

The author A.A. Bolonkin originated the AB Criterion, which allows estimating the condition required for the artificial explosion of the Sun.

11. Discussion

If we (humanity—unfortunately in this context, an insane dictator representing humanity for us) create a zone of limited size with a high temperature capable of overcoming the Coulomb barrier (for example by insertion of a specialized thermonuclear warhead) into the solar photosphere (or lower), can this zone ignite the Sun's photosphere (ignite the Sun's full load of thermonuclear fuel)? Can this zone self-support progressive runaway reaction propagation for a significant proportion of the available thermonuclear fuel?

If it is possible, researchers can investigate the problems: What will be the new solar temperature? Will this be metastable, decay or runaway? How long will the transformed Sun live, if only a minor change? What the conditions will be on the Earth during the interval, if only temporary? If not an explosion but an enhanced burn results the Sun might radically increase in luminosity for –say--a few hundred years. This would suffice for an average Earth temperature of hundreds of degrees over 0 °C. The oceans would evaporate and Earth would bake in a Venus like greenhouse, or even lose its' atmosphere entirely.

It would not take a full scale solar explosion, to annihilate the Earth as a planet for Man. (For a classic report on what makes a planet habitable, co-authored by Issac Asimov, see
http://www.rand.org/pubs/commercial_books/2007/RAND_CB179-1.pdf .

Converting the sun even temporarily into a 'superflare' star, (which may hugely vary its output by many percent, even many times) over very short intervals, not merely in heat but in powerful bursts of shorter wavelengths) could kill by many ways, notably ozone depletion—thermal stress and atmospheric changes and hundreds of others of possible scenarios—in many of them, human civilization would be annihilated. And in many more, humanity as a species would come to an end.

Fig. 4. Sun explosion

Fig. 5. Sun explosion. Result on the Earth.

The reader naturally asks: Why even contemplate such a horrible scenario? It is necessary because as thermonuclear and space technology spreads to even the least powerful nations in the centuries ahead, a dying dictator having thermonuclear missile weapons can produce (with some considerable mobilization of his military/industrial complex)— the artificial explosion of the Sun and take into his grave the whole of humanity. It might take tens of thousands of people to make and launch the hardware, but only a very few need know the final targeting data of what might be otherwise a weapon purely thought of (within the dictator's defense industry) as being built for peaceful, deterrent use.

Those concerned about Man's future must know about this possibility and create some protective system— or ascertain on theoretical grounds that it is entirely impossible, which would be comforting.

Suppose, however that some variation of the following is possible, as determined by other researchers with access to good supercomputer simulation teams. What, then is to be done?

The action proposed depends on what is shown to be possible.

Suppose that no such reaction is possible—it dampens out unnoticeably in the solar background, just as no fission bomb triggered fusion of the deuterium in the oceans proved to be possible in the Bikini test of 1946. This would be the happiest outcome.

Suppose that an irruption of the Sun's upper layers enough to cause something operationally similar to a targeted 'coronal mass ejection' – CME-- of huge size targeted at Earth or another planet? Such a CME like weapon could have the effect of a huge electromagnetic pulse. Those interested should look up data on the 1859 solar superstorm, the Carrington event, and the Stewart Super Flare. Such a CME/EMP weapon might target one hemisphere while leaving the other intact as the world turns. Such a disaster could be surpassed by another step up the escalation ladder-- by a huge hemisphere killing thermal event of ~12 hours duration such as postulated by science fiction writer Larry Niven in his 1971 story "Inconstant Moon"—apparently based on the Thomas Gold theory (ca. 1969-70) of rare solar superflares of 100 times normal luminosity. Subsequent research[18] (Wdowczyk and Wolfendale, 1977) postulated horrific levels of solar activity, ozone depletion and other such consequences might cause mass extinctions. Such an improbable event might not occur naturally, but could it be triggered by an interested party? A triplet of satellites monitoring at all times both the sun from Earth orbit and the 'far side' of the Sun from Earth would be a good investment both scientifically and for purposes of making sure no 'creative' souls were conducting trial CME eruption tests!.

Might there be peaceful uses for such a capability? In the extremely hypothetical case that a yet greater super-scale CME could be triggered towards a given target in space, such a pulse of denser than naturally possible gas might be captured by a giant braking array designed for such a purpose to provide huge stocks of hydrogen and helium at an asteroid or moon lacking these materials for purposes of future colonization.

A worse weapon on the scale we postulate might be an asymmetric eruption (a form of directed thermonuclear blast using solar hydrogen as thermonuclear fuel), which shoots out a coherent (in the sense of remaining together) burst of plasma at a given target without going runaway and consuming the outer layers of the Sun. If this quite unlikely capability were possible at all (dispersion issues argue against it—but before CMEs were discovered, they too would have seemed unlikely), such an apocalyptic 'demo' would certainly be sufficient emphasis on a threat, or a means of warfare against a colonized solar system. With a sufficient thermonuclear burn –and if the condition of nondispersion is fulfilled—might it be possible to literally strip a planet—Venus, say—of its' atmosphere? (It might require a mass of fusion fuel— and a hugely greater non-fused expelled mass comparable in total to the mass to be stripped away on the target planet.) .

It is not beyond the limit of extreme speculation to imagine an expulsion of this order sufficient to strip Jupiter's gas layers off the 'Super-Earth' within. —To strip away 90% or more of Jupiter's mass (which otherwise would take perhaps ~400 Earth years of total solar output to disassemble with perfect efficiency and neglecting waste heat issues). It would probably waste a couple Jupiter masses of material (dispersed hydrogen and helium). It would be an amazing engineering capability for long term space colonization, enabling substantial uses of materials otherwise unobtainable in nearly all scenarios of long term space civilization.

Moving up on the energy scale-- "boosting" or "damping" a star, pushing it into a new metastable state of greater or lesser energy output for times not short compared with the history of civilization, might be a very welcome capability to colonize another star system—and a terrifying reason to have to make the trip.

And of course, in the uncontrollable case of an induced star explosion, in a barren star system it could provide a nebula for massive mining of materials to some future super-civilization. It is worth noting in this connection that the Sun constitutes 99.86 percent of the material in the Solar System, and Jupiter another .1 percent. Literally a thousand Earth masses of solid (iron, carbon) building materials might be possible, as well as thousands of oceans of water to put inside space colonies in some as yet barren star system.

But here in the short-term future, in our home solar system, such a capability would present a terrible threat to the survival of humanity, which could make our own solar system completely barren.

The list of possible countermeasures does not inspire confidence. A way to interfere with the reaction (dampen it once it starts)? It depends on the spread time, but seems most improbable. We cannot even stop nuclear reactions once they take hold on Earth—the time scales are too short.

Is defense of the Sun possible? Unlikely—such a task makes missile defense of the Earth look easy. Once a gravity assist Jupiter flyby nearly stills the velocity with which a flight article orbits the Sun, it will hang relatively motionless in space and then begin the long fall to fiery doom. A rough estimate yields only one or two weeks to intercept it within the orbit of Mercury, and the farther it falls the faster it goes, to science fiction-like velocities sufficient to reach Pluto in under six weeks before it hits.

A perimeter defense around the Sun? The idea seems impractical with near term technology.

The Sun is a hundred times bigger sphere than Earth in every dimension. If we have 10,000 ready to go interceptor satellites with extreme sunshields that function a few solar radii out each one must be able to intercept with 99% probability the brightening light heading toward its' sector of the Sun over a circle the size of Earth, an incoming warhead at around 600 km/sec.

If practical radar range from a small set is considered (4th power decline of echo and return) as 40,000 km then only 66 seconds would be available to plot a firing solution and arm for a destruct attempt. More time would be available by a telescope looking up for brightening, infalling objects—but there are many natural incoming objects such as meteors, comets, etc. A radar might be needed just to confirm the artificial nature of the in-falling object (given the short actuation time and the limitations of rapid storable rocket delta-v some form of directed nuclear charge might be the only feasible countermeasure) and any leader would be reluctant to authorize dozens of nuclear explosions per year automatically (there would be no time to consult with Earth, eight light-minutes away—and eight more back, plus decision time). But the cost of such a system, the reliability required to function endlessly in an area in which there can presumably be no human visits and the price of its' failure, staggers the mind. And such a 'thin' system would be not difficult to defeat by a competent aggressor…

A satellite system near Earth for destroying the rockets moving to the Sun may be a better solution, but with more complications, especially since it would by definition also constitute an effective missile defense and space blockade. Its' very presence may help spark a war. Or if only partially complete but under construction, it may invite preemption, perhaps on the insane scale that we here discuss…

Astronomers see the explosion of stars. They name these stars novae and supernovae—"New Stars" and try to explain (correctly, we are sure, in nearly all cases) their explosion by natural causes. But some few of them, from unlikely spectral classifications, may be result of war between civilizations or fanatic dictators inflicting their final indignity upon those living on planets of the given star. We have enough disturbed people, some in positions of influence in their respective nations and organizations and suicide oriented violent people on Earth. But a nuclear bomb can destroy only one city. A dictator having possibility to destroy the Solar System as well as Earth can blackmail all countries—even those of a future Kardashev scale 2 star-system wide civilization-- and dictate his will/demands on any civilized country and government. It would be the reign of the crazy over the sane.

Author A.A. Bolonkin already warned about this possibility in 2007 (see his interview http://www.pravda.ru/science/planet/space/05-01-2007/208894-sun_detonation-0 [15] (in Russian) (A translation of this is appended at the end of this article) and called upon scientists and governments to research and develop defenses against this possibility. But some people think the artificial explosion of Sun impossible. This led to this current research to give the conditions where such detonations are indeed possible. That shows that is conceivably possible even at the present time using current rockets and nuclear bombs—and only more so as the centuries pass. Let us take heed, and know the risks we face—or disprove them.

The first information about this work was published in [15]. This work produced the active Internet discussion in [19]. Among the raised questions were the following:

1) It is very difficult to deliver a warhead to the Sun. The Earth moves relative to the Sun with a orbital velocity of 30 km/s, and this speed should be cancelled to fall to the Sun. Current rockets do not suffice, and it is necessary to use gravitational maneuvers around planets. For this reason (high delta-V (velocity changes required) for close solar encounters, the planet Mercury is so badly investigated (probes there are expensive to send).

Answer: The Earth has a speed of 29 km/s around the Sun and an escape velocity of only 11 km/s. But Jupiter has an orbital velocity of only 13 km/sec and an escape velocity of 59.2 km/s. Thus, the gravity assist Jupiter can provide is more than the Earth can provide, and the required delta-v at that distance from the Sun far less—enough to entirely cancel the sun-orbiting velocity around the Sun, and let it begin the long plunge to the Solar orb at terminal velocity achieving Sun escape speed 617.6 km/s. Notice that for many space exploration maneuvers, we require a flyby of Jupiter, exactly to achieve such a gravity assist, so simply guarding against direct launches to the Sun from Earth would be futile!

2) Solar radiation will destroy any a probe on approach to the Sun or in the upper layers of its photosphere.

Answer: It is easily shown, the high efficiency AB-reflector can full protection the apparatus. See [7] Chapters 12, 3A, [8] Ch.5, [9]-[12].

3) The hydrogen density in the upper layers of the photosphere of the Sun is insignificant, and it would be much easier to ignite hydrogen at Earth oceans if it in general is possible.

Answer: The hydrogen density is enough known. The Sun has gigantic advantage – that is PLASMA. Plasma of sufficient density reflects or blocks radiation—it has opacity. That means: **no radiation losses in detonation**. It is very important for heating. The AB Criterion in this paper is received for PLASMA. Other planets of Solar system have MOLECULAR atmospheres which passes radiation. No sufficient heating – no detonation! The water has higher density, but water passes the high radiation (for example γ-radiation) and contains a lot of oxygen (89%), which may be bad for the thermonuclear reaction. This problem needs more research.

12. Summary

This is only an initial investigation. Detailed supercomputer modeling which allows more accuracy would greatly aid prediction of the end results of a thermonuclear explosion on the solar photosphere.

Author invites the attention of scientific society to detailed research of this problem and devising of protection systems if it proves a feasible danger that must be taken seriously. The other related ideas author Bolonkin offers in [5]-[15].

References

(The reader find some author's works in http://Bolonkin.narod.ru/p65.htm, http://www.scribd.com search "Bolonkin; http://Arxiv.org Search: "Bolonkin", in http://aiaa.org search "Bolonkin" and books: Bolonkin A.A., *"Non-Rocket Space Launch and flight"*, Elsevier, 2006, 488 pgs.; Bolonkin A.A., *"New Concepts, ideas, and Innovations in Technology and Human life"*, NOVA, 2008, 502 pg.; Bolonkin A.A., Cathcart R.B., *"Macro-Projects: Environment and Technology"*, NOVA, 2009, 536 pgs).

1. AIP Physics desk reference, 3rd Ed., Spring, 888 pgs.
2. Handbook of Physical Quantities, Ed. Igor Grigoriev, CRC Press, 1997, USA.
3. I.K. Kikoin (Ed.), Tables of physical values, Atomizdat, Moscow, 1975, 1006 pgs, (in Russian).
4. Nishikawa K., Wakatani M., Plasma Physics, Spring, 2000.
5. Bolonkin A.A., New AB-Thermonuclear Reactor for Aerospace, Presented as AIAA-2006-7225 to Space-2006 Conference, 19-21 September, 2006, San Jose, CA, USA (see also http://arxiv.org search "Bolonkin").
http://arxiv.org/ftp/arxiv/papers/0706/0706.2182.pdf ,
http://arxiv.org/ftp/arxiv/papers/0803/0803.3776.pdf .

6. Bolonkin A.A., Simplest AB-Thermonuclear Space Propulsion and Electric Generator,
http://arxiv.org search "Bolonkin". http://arxiv.org/ftp/physics/papers/0701/0701226.pdf .

7. Bolonkin A.A., *"Non-Rocket Space Launch and Flight"*, Elsevier, 2006, 488 pgs.
http://Bolonkin.narod.ru/p65.htm , or http://www.scribd.com/doc/24056182 . The book contains theories of the more then 20 new revolutionary author ideas in space and technology.

8. Bolonkin A.A., *New concepts, ideas and innovations in aerospace and technology*, Nova, 2007. The book contains theories of the more then 20 new revolutionary author ideas in space and technology. http://Bolonkin.narod.ru/p65.htm , or http://www.scribd.com/doc/24057071 .
9. Bolonkin A.A., Cathcart R.B., *"Macro-Projects: Environment and Technology"*, NOVA, 2009, 536 pgs. http://Bolonkin.narod.ru/p65.htm . http://www.scribd.com/doc/24057930 . Book contains many new revolutionary ideas and projects.
10. Bolonkin A.A., High Speed AB Solar Sail. This work is presented as paper AIAA-2006-4806 for 42 Joint Propulsion Conference, Sacramento, USA, 9-12 July, 2006, USA (see also http://arxiv.org search "Bolonkin").http://arxiv.org/ftp/physics/papers/0701/0701073.pdf .
11. Bolonkin A.A., Light Multi-reflex Engine, Journal of British Interplanetary Society, Vol 57, No.9/10, 2004, pp. 353-359.
12. Bolonkin, A.A., Light Pressure Engine, Patent (Author Certificate) # 1183421, 1985, USSR

    (priority on 5 January 1983).

13. Bolonkin A.A., Converting of Matter to Nuclear Energy by AB-Generator. *American Journal of Enginering and Applied Sciences.* 2 (2), 2009, p.683-693. [on line] http://www.scipub.org/fulltext/ajeas/ajeas24683-693.pdf or http://www.scribd.com/doc/24048466/

14. Bolonkin A.A., Femtotechnology. Nuclear AB-Matter with Fantastic Properties, *American Journal of Enginering and Applied Sciences.* 2 (2), 2009, p.501-514. [On line]: http://www.scipub.org/fulltext/ajeas/ajeas22501-514.pdf or http://www.scribd.com/doc/24046679/

15. Bolonkin A.A., Artificial Explosion of Sun. Interview for newspaper www.PravdaRu.ru of 5

    January 2007. http://www.pravda.ru/science/planet/space/05-01-2007/208894-sun_detonation-0 (in Russian).

16. Wikipedia. Some background material in this article is gathered from Wikipedia under the Creative Commons license. http://wikipedia.org .
17. Solar Physics Group at NASA's Marshall Space Flight Center website for solar facts http://solarscience.msfc.nasa.gov/
18. Wdowczyk J and Wolfendale A W, Cosmic rays and ancient catastrophes, Nature, 268 (1977) 510. Abstract available at: http://www.nature.com/nature/journal/v268/n5620/abs/268510a0.html
19. Turchin A.V., The possibility of artificial fusion explosion of giant planets and other objects of Solar system, 2009. http://www.scribd.com/doc/8299748/Giant-planets-ignition.

Appendix

Here there are values useful for calculations and estimations of macro-projects.

1. System of Mechanical and Electrical Units

The following table contains the delivered metric mechanical and the electromagnetic SI units that have been introduced in this text, expressed in terms of the fundamental units *meter*, *kilogram*, *second*, and *ampere*. From these expressions the *dimensions* of the physical quantities involved can be readily determined.

Length.................. 1 meter = 1 m
Mass................. ....1 kilogram = 1 kg
Time................. .... 1 second = 1 s
Electric current.... 1 ampere = 1 A

Force......... 1 newton = 1 N = 1 kg·m/s^2
Pressure.....1 N/m2 = 1 kg/m·s^2
Energy........ 1 joule = 1 J =1 N/m = 1 kg·m^2/s^2
Power........ 1 watt = 1 W =1 J/s =1 kg·m^2/s^3

Rotational inertia.................1 kilogram·meter^2 = 1 kg·m^2
Torque............................1 meter·newton = 1 kg·m^2/s^2
Electric charge....................1 coulomb = 1 C = 1 A·s
Electric intensity..................1 N/C = 1 V/m = 1 kg·m/s^3·A

Electric potential..................1 volt = 1 V = 1 J/C = 1 kg·m^2/s^3·A
Electric resistance................1 ohm =1 Ω =1 V/A = 1 kg·m^2/s^3·A^2
Capacitance....................... 1 farad = 1 F =1 C/V = 1 C2/J = 1 s^4·A^2/kg·m^2
Inductance......................... 1 henry = 1 H =1 J/A^2 = 1 Ω·s = 1 kg·m^2/s^2·A^2

Magnetic flux......................1 webwer = 1 Wb = 1 J/A = 1 V.s = 1 kg·m^2/s^2·A
Magnetic intensity............ 1 tesla = 1 Wb/m2 =1 V.s/m2 =1 kg/s^2·A=N/mA
Reluctance......................... 1 ampera-turn/weber = 1 A/Wb =1 s^2·A^2/kg·m^2
Magnetizing force..............1 ampere-turn/meter = 1 A/m

Kelvin is fundamental unit of temperature
Candela is fundamental power-like unit of photometry

Fundamental Physical Constants

Speed of light in vacuum	c = 299 792 459 ~ 3×10^8 m/s
Magnetic constant (permeability)	μ_o = 4π×10^{-7} N/A^2
Electric constant $1/\mu_o c^2$	ε_o = 8.854 187 817...×10^{-12} F/m
Plank constant	h = 6.626 068 76...×10^{-34} J s
	$h/2\pi$ = 1.054 571 596...×10^{-34} J s
Standard gravitational acceleration	9.806 65 m/s^2
Standard atmosphere (atm)	101 325 N/m^2
Thermochemical kilocalorie	4184 J
Speed of light in vacuum (c)	2.997 935×10^8 m/s
Electronic charge (e)	1.60210×10^{-19} C
Avogadro constant (N_A)	6.0225×10^{26}/kmol

Faraday constant (F)	9.6487×10^7 C/kmol	
Universal gas constant (R)	8314 J/kmol	
Gravitational constant (G)	6.67×10^{-11} N·m^2/kg^2	
Boltzmann constant (k)	1.3806×10^{-23} J/K	
Stefan-Boltzmann Constant (σ)	5.670×10^{-8} W/K^4·m^2	
Rest energy of one atomic mass unit	931.48 MeV	
Electron-volt (eV)	1.60218×10^{-19} J	

Rest masses of particles

	(u)	(kg)	(MeV)
Electron	5.48597×10^{-4}	9.1091×10^{-31}	0.511 006
Proton	1.002 2766	1.67252×10^{-27}	938.26
α-particles	4.001 553	6.6441×10^{-27}	3727.3

Detonation energy of 1 kiloton of high explosive is 10^{12} cal. 1 cal = 4.19 J.

Universe (2016)

Age	13.799 ± 0.021 billion years
Diameter	At least 91 billion light-years (28 billion parsecs)
Mass (ordinary matter)	At least 10^{53} kg
Average density	4.5×10^{-31} g/cm^3
Average temperature	2.72548 K

Main Contents:
- Ordinary (baryonic) matter (4.9%)
- Dark matter (26.8%)
- Dark energy (68.3%)

Shape Flat with only a 0.4% margin of error

Sun

Volume	1.41×10^{18} km^3
Mass	$1.98855 \pm 0.00025) \times 10^{30}$ kg
Average density	1.408 g/cm^3
Center density (modeled)	162.2 g/cm^3
Equatorial surface gravity	274.0 m/s^2

Short biography of Bolonkin, Alexander Alexandrovich (1933-)

Alexander A. Bolonkin was born in the former USSR. He holds doctoral degree in aviation engineering from Moscow Aviation Institute and a post-doctoral degree in aerospace engineering from Leningrad Polytechnic University. He has held the positions of senior engineer in the Antonov Aircraft Design Company and Chairman of the Reliability Department in the Clushko Rocket Design Company. He has also lectured at the Moscow Aviation Universities. Following his arrival in the United States in 1988, he lectured at the New Jersey Institute of Technology and worked as a Senior Scientist at NASA and the US Air Force Research Laboratories. Home page of author is: http://Bolonkin.narod.ru .

Given book contains researches five new ideas: new preon interaction theory of the micro World; relations between time, mass, space, charge and energy; possibility of creating the super-strong (in millions times) matter, having suprice properties; super-strong nuclear AB-needles, which allows to penetrate deep into the Earth and planets; the nuclear geterator that is converting of any matter into energy.

1. In Chapter 1 the author offers some initial ideas about a cognitive construct of the Micro-World with allows to design a preon based Universe matching many qualities of the observable Universe. The main idea is that - the initial base must be very simple: two energy massless virtual particles (eners) and two reciprocity relations (interactions) between them. Author postulates: Two energy massless virtual particles can explain the main features of much of what we see including: mass, electrical charges and the main interactions between particles such as: gravitation, centrifugal and inertial masses, repulsion and attraction of electric charges, weak and strong nuclear forces, design of quarks and baryonic matter.

2. In Chapter 2 author has developed a theory which allows derivation of the unknown relations between the main parameters (energy, time, volume, matter) in the Universe. In given part 3 he added charge as main parameter in this theory. He finds also the quantum (minimal values) of energy, time, volume and matter and he applied these quantum for estimations of quantum volatility and the estimation of some values of our Universe and received both well-known and new unknown relations.

Author offers possibly valid relations between charge, time, matter, volume, distance, and energy. The net picture derived is that in the Universe exists ONLY one substance – ENERGY. Charge, time, matter, volume, fields are evidence of this energy and they can be transformed one to other. Author gives the equations which allow to calculate these transformation like the famous formula $E = mc^2$. Some assumptions about the structure of the Universe follow from these relations.

3-4. In Chapter 3-4 the author researches the design the super-strong matter. This matter is stronger than convetional mathriales in millions times. It is can withstand temperatures in millions degree. Aerospace, aviation particularly need, in any era, the strongest and most thermostable materials available, often at nearly any price. The Space Elevator, space ships (especially during atmospheric reentry), rocket combustion chambers, thermally challenged engine surfaces, hypersonic aircraft materials better than any now available, with undreamed of performance as the reward if obtained. As it is shown in this research, the offered new material allows greatly to improve the all characteristics of space ships, rockets, engines and aircraft and design new types space, propulsion, aviation systems.

5. In Chapter 5 Author offers a new nuclear generator which allows to convert any matter to nuclear energy in accordance with the Einstein equation $E=mc^2$.

6. In Chapter 6 the author is exploring the possibility of artificial sun explosion.

www.ingramcontent.com/pod-product-compliance
Lightning Source LLC
Chambersburg PA
CBHW080939170526
45158CB00008B/2307